Problems of empiricism

Philosophical papers
Volume 2

PAUL K. FEYERABEND

Published by the Press Syndicate of the University of Cambridge
The Pitt Building, Trumpington Street, Cambridge CB2 1RP
40 West 20th Street, New York, NY 10011-4211, USA
10 Stamford Road, Oakleigh, Melbourne 3166, Australia

First published 1981
First paperback edition 1985
Reprinted 1986, 1988, 1989, 1994, 1995

Library of Congress Cataloging-in-Publication Data is available.

A catalogue record for this book is available from the British Library.

ISBN 0-521-31641-3 paperback

Transferred to digital printing 2004

Problems of empiricism

Contents

Introduction to volumes 1 and 2

The present volume and its companion discuss three ideas that have played an important role in the history of science, philosophy and civilization: criticism, proliferation and reality. The ideas are presented, explained and made the starting points of argumentative chains.

The first idea, that of *criticism*, is found in almost all civilizations. It plays an important role in philosophies such as Buddhism and Mysticism, it is the cornerstone of late nineteenth-century science and philosophy of science, and it has been applied to the theatre by Diderot and Brecht.[1] Criticism means that we do not simply accept the phenomena, processes, institutions that surround us but we examine them and try to change them. Criticism is facilitated by *proliferation* (vol. 1, ch. 8): we do not work with a single theory, system of thought, institutional framework until circumstances force us to modify it or to give it up; we use a plurality of theories (systems of thought, institutional frameworks) from the very beginning. The theories (systems of thought, forms of life, frameworks) are used in their strongest form, not as schemes for the processing of events whose nature is determined by other considerations, but as accounts or determinants of this very nature (*realism*, see vol. 1, chs. 11.15f[2]). One chain of argument is therefore

$$\text{criticism} \Rightarrow \text{proliferation} \Rightarrow \text{realism} \qquad \text{(i)}$$

In the first volume this chain is applied to a rather narrow and technical problem, viz. the interpretation of *scientific theories*.

None of the ideas is defined in a precise fashion. This is quite intentional. For although some papers, especially the early ones, are fairly abstract and 'philosophical', they still try to stay close to scientific practice which means that their concepts try to preserve the fruitful imprecision of this practice (cf. vol. 2, ch. 5 on the ways of the scientist and the ways of the philosopher; cf. also vol. 2, ch. 6, nn. 47ff and text).

Nor does the arrow in (i) express a well-defined connection such as

[1] This wider function of criticism is explained in my essay 'On the Improvement of the Sciences and the Arts and the Possible Identity of the Two' in *Boston Studies in the Philosophy of Science* (New York, 1965), III.

[2] ch. 11.15 means section 15 of ch. 11. This method of reference is used throughout both volumes.

logical implication. It rather suggests that starting with the left hand side and adding physical principles, psychological assumptions, plausible cosmological conjectures, absurd guesses and plain commonsense views, a dialectical debate will eventually arrive at the right hand side. Examples are the arguments for proliferation in vol. 1, ch. 6.1, ch. 8, n.14 and text, ch. 4.6 as well as the arguments for realism in vol. 1, chs. 11, 14 and 15. The meaning of the arrow emerges from these examples and not from example-independent attempts at 'clarification'.

Chapters 2–7 of vol. 1, which are some of the oldest papers, deal mainly with the interpretation of theories (for the notion of 'theory' used cf. the remarks in the preceding paragraph and in vol. 1, ch. 6, n.5). Chapter 1 of the first volume shows how the realism that is asserted in thesis 1 of vol. 1, ch. 2.6 and again in ch. 11, is related to other types of realism that have been discussed by scientists. The thesis can be read as a philosophical thesis about the influence of theories on our observations. It then asserts that observations (observation terms) are not merely theory-*laden* (the position of Hanson, Hesse and others) but *fully theoretical* (observation statements have no 'observational core'[3]). But the thesis can also be read as a historical thesis concerning the use of theoretical terms by scientists. In this case it asserts that scientists often use theories to restructure abstract matters *as well as* phenomena, and that no part of the phenomena is exempt from the possibility of being restructured in this way. My discussion of the relation between impetus and momentum in vol. 1, ch. 4.5 is entirely of the second kind. It is not an attempt to draw consequences from a contextual theory of meaning – theories of meaning play no role in this discussion – it simply shows that both facts and the laws of Newtonian mechanics prevent us from using the concept of impetus as part of Newton's theory of motion. Nor is the result generalized to all competing theories. It is merely argued that certain popular views on explanation and the relation between theories in the same domain that claim to be universally valid fail for important scientific developments. *General* assertions about incommensurability are more characteristic for Kuhn whose ideas differ from mine and were developed independently (cf. my *Science in a Free Society*,[4] 65ff for a comparison and a

[3] Or, to express it differently: there are only theoretical terms (for his version of the thesis see my 'Das Problem der Existenz theoretischer Entitäten' in *Probleme der Wissenschaftstheorie*, ed. E. Topitsch (Vienna, 1960), 35ff). There is of course a distinction between theoretical terms and observation terms, but it is a psychological distinction, dealing with the psychological processes that accompany their use, but having nothing to do with their content (for details see vol. 1, ch. 6, section 6). This feature of the thesis has been overlooked by some more recent critics who ascribed to me the 'triviality that theoretical terms are theoretical'. The best and most concise expression of the thesis can be found in Goethe: 'Das Hoechste zu begreifen waere, dass alles Faktische schon Theorie ist' ('Aus den Wanderjahren', *Insel Werkausgabe* (Frankfurt, 1970), vi 468). [4] (London, 1978), hereafter referred to as *SFS*.

brief history. Chapter 17 of my *Against Method*[5] discusses a special case that shows what elements must be considered in any detailed discussion of incommensurability). There do exist cases where not only do *some* older concepts break the framework of a new theory, but where *an entire theory*, all its observation statements included, is incommensurable with the theory that succeeds it, but such cases are rare and need special analysis. Using the terms of vol. 1, ch. 2.2 one can tentatively say that a theory is incommensurable with another theory if its ontological consequences are incompatible with the ontological consequences of the latter (cf. also the considerations in vol. 1, ch. 4.7 as well as the more concrete definition in *A M*, 269 and the appendix to ch. 8 of vol. 2). But even in this case incommensurability does not lead to complete disjointness, as the phenomenon depends on a rather subtle connection between the more subterranean machinery of the two theories (cf. again *A M*, 269). Besides there are many ways of comparing incommensurable frameworks, and scientists make full use of them (vol. 1, ch. 1, n.39; cf. also vol. 1, ch. 2.6, n.21 and ch. 4.8). Incommensurability is a difficulty for some rather simpleminded philosophical views (on explanation, verisimilitude, progress in terms of content increase); it shows that these views fail when applied to scientific practice; it does not create any difficulty for scientific practice itself (see vol. 2, ch. 11.2, comments on incommensurability).

Chapters 8–15 of vol. 1 apply chain (1) to the mind–body problem, commonsense, the problem of induction, far-reaching changes in outlook such as the Copernican revolution and the quantum theory. The procedure is always the same: attempts to retain well-entrenched conceptions are criticized by pointing out that the excellence of a view can be asserted only *after* alternatives have been given a chance, that the process of knowledge acquisition and knowledge improvement must be kept in motion and that even the most familiar practices and the most evident forms of thought are not strong enough to deflect it from its path. The cosmologies and forms of life that are used as alternatives need not be newly invented; they may be parts of older traditions that were pushed aside by overly eager inventors of New Things. The whole history is mobilized in probing what is plausible, well established and generally accepted (vol. 1, ch. 4, n. 67 and ch. 6.1).

There is much to be said in favour of a pluralistic realism of this kind. John Stuart Mill has explained the arguments in his immortal essay *On Liberty* which is still the best modern exposition and defence of a critical philosophy (see vol. 1, ch. 8 and vol. 2, ch. 4 and ch. 9.13). *But the drawbacks are considerable.* To start with, modern philosophers of science, 'critical' rationalists included, base their arguments on only a tiny part of Mill's scheme; they uncritically adopt some standards, which they use for weeding

[5] (London, 1975), hereafter referred to as *A M*.

out conflicting ideas, but they hardly ever examine the standards themselves. Secondly, and much more importantly, there may be excellent reasons for resisting the universal application of the realism of thesis 1 (vol. 1, ch. 2.6). Take the case of the quantum theory. Interpreted in accordance with vol. 1, ch. 2.6 wave mechanics does not permit the existence of well-defined objects (see vol. 1, ch. 16.3 and especially nn.26, 27 and text). Commonsense (as refined by classical physics) tells us that there are such objects. The realism of vol. 1, ch. 2.6, chs. 4.6f, chs. 9 and 11 invites us to reject commonsense and to announce the discovery: objective reality has been found to be a metaphysical mistake.

Physicists did not go that way, however. They demanded that some fundamental properties of commonsense be preserved and so they either added a further postulate (reduction of the wave packet) leading to the desired result, or constructed a 'generalized quantum theory' whose propositions no longer form an irreducible atomic lattice. The proponents of hidden variables, too, want to retain some features of the classical (commonsense) level and they propose to change the theory accordingly. In all these cases (excepting, perhaps, the last) a realistic interpretation of the quantum theory is replaced by a partial instrumentalism.

Two elements are contained in this procedure and they are not always clearly separated. The first element which affects the actions of the physicists is *factual*: *there are* relatively isolated objects in the world and physics must be capable of describing them. (Commonsense arguments, though more complex, often boil down to the same assertion.) But Buddhist exercises create an experience that no longer contains the customary distinctions between subject and object on the one hand and distinct objects on the other. Indeed many philosophies deny separate existence and regard it as illusion only: the existence of separate objects and the experiences confirming it are not tradition-independent 'facts'; they are parts of special traditions. Physicists *choose* one of these traditions (without realizing that a choice is being made) and turn it into a boundary condition of research. This is the second element. The transition to a partial instrumentalism therefore consists of a *choice* and the utilization of the *facts* that belong to the tradition chosen.

The history of philosophy offers many cases for the study of both elements. The debates about the quantum theory have much in common with the ancient issue between Parmenides (and his followers) on the one hand and Aristotle on the other and the more recent issue between Reason and the (Roman) Church. Parmenides showed (with additional arguments provided by Melissus and Zeno) that there is no change and that Being has no parts. But we deal with change and with objects and processes that differ in many respects. Our lives as human beings are directed towards taking change and division into account. Are we to admit that we live an illusion, that the truth is hidden from us and that it must be discovered by special means? Or should we not rather assert the reality of our common views over the reality of some specialist conceptions? Must we adapt our lives to the ideas and rules devised by small groups of intellectuals (physicians, medical researchers, socio-biologists, 'rationalists' of all sorts) or should we not rather demand that intellectuals be mindful of circumstances that matter to their fellow human beings? Or, to consider a more important dichotomy: can we regard our lives on this earth and the ideas we have developed to cope with the accidents we encounter as measures of reality, or are they of only secondary importance when compared with the conditions of the soul as described in religious beliefs? These are the questions which arise when we compare commonsense with religious notions or with the abstract ideas that intellectuals have tried to put over on us ever since the so-called rise of rationalism in the West (see vol. 2, ch. 1, sections 1f and 7). They involve both a choice between forms of life and an adaptation of our ideas and habits to the ideas (perceptions, intuitions) of the tradition chosen: *we decide to regard those things as real which play an important role in the kind of life we prefer.*

Making the decision we start a reverse argumentative chain of the form

$$L \Rightarrow \overline{\text{criticism}} \Rightarrow \overline{\text{realism}}_{\bar{L}} \qquad \text{(ii)}$$

or, in words: accepting a form of life L we reject a universal criticism and the realistic interpretation of theories not in agreement with L. Proceeding in this way we notice that instrumentalism is not a philosophy of defeat; it is often the result of far-reaching ethical and political decisions. Realism, on the other hand, only reflects the wish of certain groups to have their ideas accepted as the foundations of an entire civilization and even of life itself.

Chapters 16 and 17 of vol. 1 as well as the essays in vol. 2 contain first steps towards undermining this intellectual arrogance. They contain instances of the use of the reverse chain (ii). It is argued that science never obeys, and cannot be made to obey, stable and research independent standards (vol. 2, ch. 1.5, chs. 8, 10, 11): scientific standards are subjected to the process of research just as scientific theories are subjected to that process (vol. 1, chs. 1.3f; cf. also part 1 of *SFS*); they do not guide the process from the outside (cf. vol. 2, ch. 7 on rules and vol. 2, ch. 5 on the difference between the scientists' way and the philosophers' way of solving problems). It is also shown that philosophers of science who tried to understand and to tame science with the help of standards and methodologies that transcend research, *have failed* (vol. 2, chs. 9, 10, 11; cf. vol. 2, ch. 1.5f and part 1 of *SFS*): *one of the most important and influential institutions of our times is beyond the reach of reason as interpreted by most contemporary rationalists.* The failure does not put an end to our attempts to adapt science to our favourite forms of life. Quite the contrary: it frees the attempt from irrelevant restrictions. This is in perfect agreement with the Aristotelian philosophy which also limits science by reference to commonsense except that conceptions of an individual philosopher (Aristotle) are now replaced by the *political decisions* emerging from the institutions of a free society (vol. 2, ch. 1.7).

Most articles were written with support from the National Science Foundation; some articles were written while I held a Humanities Research Fellowship and a Humanities Research Professorship at the University of California at Berkeley. The reader will notice that some articles defend ideas which are attacked in others. This reflects my belief (which seems to have been held by Protagoras) that good arguments can be found for the opposite sides of any issue. It is also connected with my 'development' (details in *SFS*, 107ff). I have occasionally made extensive changes both in the text and in the footnotes but I have not always given the place and nature of such changes. Chapters 1 and 8 of vol. 1 and ch. 1 of vol. 2 are new and prepare a longer case study of the rise of rationalism in the West and its drawbacks. An account of, and arguments for, my present position on the structure and authority of science can be found in vol. 2 of *Versuchungen*, ed. H. P. Dürr (Frankfurt, 1981) which contains essays by various authors commenting on and criticizing my earlier views on these matters.

I

Historical background

Some observations on the decay of the philosophy of science

1. COMMONSENSE AND ABSTRACT PHILOSOPHY

Self-contained traditions only rarely raise questions of existence and reality. A member of such a tradition may ask whether a *particular event* has occurred and he may doubt a *particular tale*, but hardly anybody considers the 'ontological implications' of *all* terms, statements, stories in a certain domain.[1]

The absence of such queries is not always a sign of simplemindedness. Quite the contrary: the critical examination of reality and existence *as a whole* usually goes hand in hand with a considerable decrease in sophistication. Commonsense views (tribal commonsense; Homeric commonsense; the use of common notions in modern Western languages) contain subtly articulated *ontologies* including spirits, dreams, rainbows, stones, animals, festivities, justice, fate, sickness, divorces, the sky, obligations, death, Watkins, fear, and so on. Each of these entities behaves in a complex and characteristic manner that cannot be made to conform to a uniform pattern or summarized in a formula (cf. ch. 7.7); each of them is related to other entities and processes constituting a rich and varied universe. In such a universe the question is not what is 'real' and what is not – queries such as these would not even count as genuine questions – the question is what occurs, in what connection, and how.

[1] This is a first approximation. The Trobriand Islanders distinguish between fairytales (which are told for entertainment), legends, historical accounts, hearsay tales (which, though different from each other, are all supposed to be true) and myths (which are not only true but also sacred): see B. Malinowski, *Magic, Science and Religion* (New York, 1954), 101ff. The Dogon use accounts known to be false to them for teaching purposes (M. Griaule, *Conversations with Ogotemmeli* (Oxford, 1965), 49), to get rid of unwelcome strangers (xiiif), or to explain their views to a foreigner, using terms and similes as close as possible to his own point of view (58). The Azande are aware of failures, but can explain them (E. E. Evans-Pritchard, *Witchcraft, Oracles and Magic with the Azande* (Oxford, 1937), 80, 337f and passim). The purpose of abstractions is easily grasped and the abstractions themselves built from the non-abstract terms of the language used (F. Boas, *The Mind of Primitive Man* (New York, 1965), 197 using American Indian dialects as examples) while the laws of abstract argument are soon learned and turned against the Western visitor (Evans-Fritchard, *The Azande*, 66f). If we consider such material it is not at all implausible to assume that traditional boundary lines between different types of stories (see the classification at the beginning of this footnote) may be shifted and that entire groups of stories may be reclassified. This may be the beginning of ontological debates in our sense. Material for these matters is difficult to find; anthropologists only rarely examine conceptual change.

Commonsense *knowledge* is equally complex. There is no separation between an 'objective' world that is what it is, no matter what is done or thought about it, and the changing beliefs of researchers who try to grasp its features. Logical and epistemic notions cut across such boundaries *and rightly so* for there are many and complex interactions between 'subject' and 'object' and many ways in which the one merges into the other.[2]

'Ontological' problems in the modern sense and corresponding 'theories of knowledge' arise when parts of a complex reality of this kind are first subsumed under abstract and often rather simpleminded concepts and are then evaluated, i.e. declared to be either 'real' or 'unreal' on that basis. They arise not because one has discovered more refined ways of thinking; they arise because delicate matters are compared with crude ideas and are found to be wanting in crudeness.[3]

One can occasionally explain why crude ideas get the upper hand. Special groups want to create a new tribal identity or to preserve an already existing tribal identity amidst a rich and varied culture; to do so they reject large parts of that culture (and of the associated reality) and regard them either as non-existent or as wholly evil. The first way was chosen by the Israelites (monotheism[4]), the second by the early Christians for whom all pagan gods became (existing) devils:[5] ontological sophistication is a luxury when the life of the tribe or of a religious community is at stake.

The rise of 'rationalism' in ancient Greece is much more difficult to explain. We have again a rich, subtle, and finely articulated cosmology – Homeric commonsense – which is gradually replaced by a crude and

[2] For a brief account see W. Schadewaldt, *Die Anfaenge der Philosophie bei den Griechen* (Frankfurt, 1979), 162ff. See also his analysis of earlier notions of truth (inspired by Heidegger) *ibid.*, 195f. Ernst Mach has criticized simpleminded conceptions of reality and indicated how scientific research can be used to change them. See chs. 5 and 6.

[3] The increase of conceptual crudeness that goes hand in hand with the rise of philosophy in the West, has often been commented upon: 'Words . . . become impoverished in content, they become one sided and empty formulae' K. von Fritz, *Philosophie und Sprachlicher Ausdruck bei Demokrit, Platon und Aristoteles* (Neudruck Darmstadt, 1966), 11. Philosophy 'wipes out the more delicate and precise distinctions of the older language' K. von Fritz, *Grundprobleme der Geschichte der Antiken Wissenschaften* (Berlin, 1971), 78. We have an 'increasing alienation . . . and a destruction of clear connections' B. Snell, *Die Ausdruecke fuer den Begriff des Wissens in der Vorplatonischen Philosophie* (Berlin, 1924), 80f. Cf. also J. J. Austin, *Sense and Sensibilia* (Oxford, 1962), 3f: 'I am not going to maintain that we ought to be "realists", to embrace, that is, the doctrine that we *do* perceive material things (or objects). This doctrine would be no less scholastic and erroneous than its antithesis. The question, do we perceive material things or sensedata, no doubt looks very simple – *too* simple – but is entirely misleading (cf. Thales' similarly vast and oversimple question, what is the world made of). There is no *one* kind of thing that we "perceive" but many *different* things, the number being reducible if at all by scientific investigation and not by philosophy.' Cf. also *AM*, ch. 17 for a more detailed description of the structure of Homeric commonsense and of the transition to 'philosophy'.

[4] Yehezekel Kaufman, *The Religion of Israel* (New York, 1972). According to Kaufman (whose ideas have been criticized by Albright) the innovator was Moses.

[5] The procedure can be traced to St Paul: 'But I say that the things which the gentiles sacrifice they sacrifice to devils and not to god.' I Corinthians, x, 20.

streamlined world. The development consists of various strands practised by different groups of people. The 'microprocesses' intersect and merge; the 'rise of rationalism' is the result.[6] Some microprocesses involve implicit and unintentional conceptual changes, others an explicit and occasionally rather naive use of abstractions. Implicit changes occur in Homer,[7] Hesiod,[8] in the Ionian historians and philosophers of nature as well as in the early history of medicine.[9] But Xenophanes, Parmenides and their followers consciously rebuild concepts and contrast their ideas with 'the opinions of the mortals',[10] i.e. with commonsense: from the very beginning intellectuals show conceit and contempt for procedures different from their own.

Explicit conceptual changes are closely connected with stories or 'arguments' told to reveal the difficulties inherent in a certain situation: S takes place; S either leads to A or requires that the action A be carried out; but A is impossible; hence we must either change S, or the way from S to A.

The plot of the *Oresteia* is an example of a story of this kind.[11] Orest's father Agamemnon has been killed (S); the death must be revenged by his son Orest (A); to carry out the revenge Orest must kill his mother which is against the law of the Eumenides. The difficulty is removed by a new social order, i.e. by a change of the 'implications' of S. A change of S itself is required by some of Zeno's arguments:[12] a finite line consists of infinitely many points (S) which are either extended, or not extended (A). If they are extended, then adding them up will produce an infinite line while adding up unextended points will never create a finite extension: A is impossible. Lines, therefore, do not consist of points.

[6] An excellent survey of the relatively independent microprocesses that constitute the phenomenon called, rather summarily, the 'rise of rationalism' is G. E. R. Lloyd, *Magic, Reason and Experience* (Cambridge, 1979). Lloyd is far ahead of those historians who are intent on isolating what they regard as 'scientific' elements and who construct lines of development from the elements they have found (or, rather, imposed). The drawback of the book lies in the fact that he regards changes towards science and rationalism as being invariably beneficial, that he underestimates their size and that he fails to look for and to identify the great and still unsolved problems created in their course. Chapter 17 of *A M* contains a preliminary account of some of these problems. I hope to produce a more detailed account in the near future.

[7] One example of such an implicit change is described in *A M*, 267.

[8] In Hesiod the transitions between the first principles satisfy a 'conservation law of qualities': chaos gives birth to Erebus and Night both of which are as dark as chaos. Earth gives birth to mountains and to the sky both of which are hard. These similarities have prompted Hans Schwabl to call Erebus and Night as 'belonging to the concept [of chaos]' (Pauly-Wissowa, article Weltschoepfung, column 8) *conceptual relations announce themselves as qualitative similarities between succeeding elements of a process of generation.*

[9] For examples cf. F. Heinimann, *Nomos und Physics* (Darmstadt, 1978; first published 1945).

[10] Parmenides, quoted from H. Diels & W. Kranz (eds.), *Die Fragmente der Vorsokratiker* I (Berlin, 1951), fragm. B1, 11ff.

[11] Historians of logic have so far failed to examine the anticipations of *modus tollens* that occur in tragedies such as the *Oresteia.* They make Reinhardt's conjecture concerning Xenophanes much more plausible. Cf. n.13.

[12] Cf. G. E. L. Owen 'Zeno and the Mathematicians' *Proc. Artist. Soc.*, (1957/58), 199ff.

The story of the *Oresteia* (which we may regard as an implicit argument) and Zeno's argument on extension have the same formal structure: if *S*, then *A* (or *A* or *B*); *A* is impossible (and *B* is impossible), therefore . . . The difference is that in the first case the story is used to throw doubt on the connection between *S* and *A* while in the second case it is used to throw doubt on *S* itself. Also by the time of Zeno the story has become standardized and the ties between its various steps have been strengthened. This is partly due to the simple nature of the entities described: divisible lines, indivisible points of finite or zero extension. The 'proof' works because the concepts that enter it allow us to bring the story to a successful conclusion. One is inclined to say that they have been specially designed to make the story succeed.

A direct influence of the hoped-for success of a story on the concepts used to tell it seems to occur in some early arguments about the nature of God.[13] We know that the popular Homeric conception of the gods was criticized both by more intellectual poets such as Euripides[14] and by early philosophers such as Xenophanes. It would seem that an important part of this criticism was the *observation* that a god who needs helpers and servants and whose actions are restricted by other gods is lacking in power. It also seems that the observation was turned into a '*proof*':[15] if there are many gods, then they are either equal or unequal. If they are unequal, then some gods rule over others; the latter, not being all powerful, are not divine. If they are equal, then they are like the inhabitants of a city, i.e. they are again not divine. But gods are divine. It 'follows' that there can be only one God.

Note the difference between the observation and the 'proof'. The observation emphasizes power over other properties; it already uses a rather uncommon notion of a deity. But while it regards a restriction of power as a *problem*, the 'proof' makes unrestricted power a *necessary condition* of being a god. It not only states the problem in a standardized way thus making the 'solution' seem inescapable; it tries to drive it home and it uses a new and abstract concept of God to bar any escape. We may conjecture that *the increasing abstraction of philosophical notions that reaches a peak in Parmenides is at least partly the result of attempts to make proof stories conclusive.*[16] Combined with the implicit tendencies to abstraction mentioned above these attempts gradually led to *new traditions* with interesting new features.

[13] Karl Reinhardt, *Parmenides* (Frankfurt, 1959), 89ff ascribes the arguments to Xenophanes. This has been contested. The idea that the arguments were pre-Parmenidean is much more difficult to refute.

[14] *Heracles*, 1341ff.

[15] Reinhardt, *Parmenides*, 95.

[16] Attributing the argument to Xenophanes, Reinhardt writes (*Parmenides*, 96, my italics): 'He tried to prove the unity of God. To carry out the proof he chose the notion of all-powerfulness: it did not occur to him that this notion was as little given as the notion of unity (neither of them were part of the popular beliefs) or at least, it did not concern him; for only the concept of unity could be grasped by him in a dialectical way – *and dialectics was his main concern.*'

Let me now explain some differences between these traditions and their commonsense predecessors.

Homeric commonsense contains complex concepts for describing delicately articulated processes and events. The concepts are rich in content but poor in similarities and, therefore, in deductive relations. Their explanation is by lists and not by common properties. This feature of the older traditions appears in the Platonic dialogues, though in a rather indirect way. Who does not remember how often Socrates, having asked for the nature of virtue, knowledge, justice receives a list and rejects it with comments such as 'I seem to be in luck – I wanted one virtue, and I find you have a whole swarm of them to offer' (*Meno*, 72a4f); or 'you are generous indeed, my dear Theaetetus – you are asked for a simple thing and you offer a whole variety' (*Theaetetus*, 146d2); and so on. Now the fact that lists turn up in almost every dialogue, that Socrates objects to them, that he occasionally encounters reasoned resistance (*Meno*, 73a2ff) and even shows an inclination to use lists himself (*Philebus*, 62b) makes us suspect that explanations using lists were not just errors, or signs of conceptual simplemindedness – although Socrates often enough treats them in this manner – but that they may have been appropriate in situations in which definitions by essence were difficult and perhaps impossible to obtain. The suspicion is correct. There exist entire traditions where processes and objects are aggregates of relatively independent parts, concepts are representations of such aggregates in different situations and where 'knowledge' enumerates what happens when we move from one situation to the next (for examples see *A M*, ch. 17). In such traditions a list is not a mistaken first step on the way to a more appropriate definition: it is the only adequate form of knowledge.

2. HISTORICAL TRADITIONS AND ABSTRACT TRADITIONS

From now on I shall call traditions with these conceptual peculiarities historical traditions. Traditions containing abstract relations between abstract concepts of relatively poor content I shall call abstract traditions. Note that lists already assume some connections between *concepts* and are therefore a step towards a more abstract account. Note also that we are dealing with open lists: they can be enriched by further items, and items contained in them can be omitted or replaced without any change in the concept that is represented by the list. There are of course limits to enrichment and change, but these limits are implicit in the use of the list; they cannot be captured in a formula; tact and not logic determine the content of a concept and the permissible changes. Conceptual matters are similar to matters of style, or of correct usage, which means they must be learned by 'immersion', as a child learns a language, and not by abstract study of abstract principles. Having learned the practice the members of a historical tradition are capable of a great variety of reactions that far exceeds any explicit account that they, or a student of their tradition, might be able to give, and the explanations they provide are far more varied than any collection of lists one might succeed in assembling at a particular time.

The fact that we learn historical traditions by immersion and not by studying formal models, that we reveal their features by using tact in concrete situations rather than by following abstract rules has important consequences for the way in which such traditions can be presented and explained. Historians of mathematics or of astronomy appear to have no problems; they introduce certain objects (principles, theorems, abstract structures that obey certain general rules) and describe how the structures were built up and how they were received. The objects of historical traditions obey not only rules but also the dictates of tact and it is very difficult to capture these dictates in objective historical descriptions (cf. ch. 7.7). Some historians therefore present their material not as *evidence* to be used for or against (historical) generalizations but as *teaching aids* for the development of the very same kind of tact that has produced the material in the first place; the reader is not an objective judge of evidence, he becomes part of the process that produces the evidence, and judges generalizations on that basis (this elucidates what some authors have said about the function of 'Verstehen' in the *Geisteswissenschaften*).

What is true of concepts is true of *arguments*. For the members of a historical tradition an argument is a story with a point. Learning to argue is part of learning how to get along with people. One does not study the 'nature of man'; one studies individual people and learns the many different ways of living with them, arguments included. One learns how to adapt one's persuasion to the idiosyncrasies of the person one confronts rather than to an abstract creature 'rational man'. There is no distinction between logic and rhetoric. The 'subjective' and the 'objective' side of an argument merge into one. Nor is it possible to 'demarcate' clearly one historical tradition from another. Notions with clearly defined intensions, such as 'science', 'myth', 'metaphysics', are incapable of capturing distinctions between elements (subtraditions) of a historical tradition.

At first sight abstract traditions seem to have entirely different properties. They are not immediately relevant to practical questions, they play hardly any role in personal relations, they disturb these relations wherever they intrude, they make them clearer, 'more rational', and impoverish them. The skills, intuitions, capabilities, emotions that are parts of a rich and rewarding life are subjected to stern controls; they lose in content and in humanity. Having been designed to facilitate the completion of certain stories or 'arguments' the concepts of these traditions can now be linked in many ways: they are poor in content but rich in deductive connections. It is interesting to see how readily these connections are used as means of enforcing consent.

I have explained that the 'arguments' of historical traditions are stories with a point. They are learned as one learns a language, or other forms of social behaviour. Different traditions and different groups within a tradition use different stories and are impressed by different 'arguments'. Individual differences force the arguer to be even more perspicacious and to pay attention to the interests, the wishes and prejudices, as well as to the 'emotional profile' of his interlocutor. Now when the Greeks discovered the great variety of forms of life in their surroundings and the many instruments of persuasion used to make them acceptable they reacted in various ways; some strengthened their own tradition (cf. the reaction of the Israelites as reported above); others mixed them with what they found and expected improvements (this is how the 'first internationalism' in the Near East seems to have arisen); still others accepted a wise and tolerant relativism which does not deny the validity of customs but makes them part of certain forms of life.[17] The discovery that abstract concepts *which are too poor to reflect the peculiarities of any particular tradition and thus seem to be tradition independent* are linked by abstract relations then suggested a further and more 'objective' way of dealing with the multitude: replace all these forms of life by a single abstract tradition, accept the 'objective' laws of this tradition and try to prove them by using the abstract relations they contain. We find this suggestion in the early philosophers, starting with Xenophanes. It has been accepted by rationalists and has become a basis of their faith.

However, the fact that the new concepts do not reflect any of the *existing* traditions does not mean that they transcend all traditions and that relativism has been conquered. Moreover, abstract traditions are not alternatives

[17] In *Herodotus* 3, 38 we find the following story: 'When Dareios was king, he had all the Greeks in his surroundings called to him and asked them what it would need to make them eat the dead bodies of their parents. However they replied they would not do it under any circumstances. Then Dareios called in the Indian Kalatians who eat the dead bodies of their parents and asked them with the Greeks present – a translator made clear to them what he said – what it would take to make them burn their dead fathers. They exclaimed in loud distress and asked him most insistently not to utter such godless words. This is what custom achieves and Pindar, in my opinion, is quite right in saying that custom is king of all beings.'

Custom is the king of all beings, but different people choose different kings: 'If one asked the peoples of the earth to choose the best customs from among those existing then each people would after a detailed examination prefer its own customs to all others – so convinced is every people that its own form of life is the best possible.' The conviction, according to Herodotos, is not nonsensical. Considering Cambyses who tore down temples and mocked customs Herodotos had this to say: 'It is completely clear to me that Cambyses was quite mad; otherwise he would not have abused temples and customs.' In sum: conviction, customs, laws are never generally accepted – but only a madman would take this as a reason for mocking them.

of historical traditions; they are special parts of them. The structures they contain, the abstract notions that enter these structures can be learned, understood, adapted to new cases only because they form parts of an underlying historical medium that supports them, gives them meaning and shows how they can be applied. Their amazing stability is the result of complex interactions in this medium or tradition; it depends on it (and on a variety of *physical* laws guaranteeing, for example, the permanence of certain *physical* signs).[18] The dichotomy between historical traditions and abstract traditions, therefore, does not reflect a real difference: *all* traditions are historical traditions (this was made very clear by the intuitionists and by Wittgenstein). But we can distinguish between empirical and theoretical traditions. *Theoretical traditions* try to replace the quasi-intuitive and only partly standardized procedures of their members by abstract models with abstract concepts and abstract relations between them and they make maximal use of these relations in their arguments. One believes that the inventions of the human mind will eventually replace the work of all our known and unknown, explicit and hidden, adaptive faculties, emotions and commonsense included, so that the whole work of creation can be built up anew, on the basis of human reason alone. More critical friends of reason concede that such a total replacement can never occur; they realize that reason can at most reform a tiny part of our natural and social surroundings and that even such a reform will only partly agree with its demands, but they still urge us to apply abstract thought whenever possible. Defenders of *empirical traditions* deny the universal usefulness of such a procedure; they assert that there are areas where theoretical traditions *can* but *should not* be introduced for empirical as well as moral reasons. In these areas reason can at most function as an *instrument* of living; it cannot be used to determine its *basic outlines*. The issue between the rationalists who favour theoretical traditions and the defenders of empirical traditions forms the background of the 'ancient battle between philosophy and poetry' (vol. 1, ch. 1, n.4), the (much less obvious) altercations between the Cartesians and the followers of Vico, the oppositions in the fourteenth century between practitioners of painting with an intuitive understanding of rules, and 'scientific' painters bent on subjecting it to a strict method, and various modern brawls such as the quarrel between scientific physicians and healers, or the quarrel between a 'rationalist', i.e. rule-bound, and a more freewheeling approach to science (cf. ch. 5). It is a permanent feature of the history of ideas.

3. HISTORIANS AND APOLOGISTS

The Platonic dialogues present the clash between a variety of empirical

[18] On this point cf. Wittgenstein's *Remarks on the Foundations of Mathematics* (Oxford, 1956), ch. 2.

traditions in cosmology, epistemology, ethics, politics, medicine, etc. and the slowly advancing 'rationalism' of the philosphers and their allies, the early scientists, as a gradual triumph of reason over ignorance and incompetence. This is obvious propaganda. However, Plato was well aware of the difficulties of a purely theoretical approach. He not only defends and praises such an approach, but also lays bare its limitations and in the end seems to adopt a compromise in which empirical traditions (empirical laws of nature and society included) have definitely the upper hand.

Almost all historians of institutions and ideas have accepted Plato's propaganda but have disregarded, or remained unaware of, his reservations. Taking the ideology of the scientists and the philosophers, i.e. taking their own 'rationalism' as a starting point, they describe any move towards it as a (long overdue) discovery or anticipation of the correct procedure, and any hesitation or return is for them a loss of valuable achievements. They are not historians; they are apologists of rationalism.[19]

To prepare a more realistic approach we must consider the following rather simple and elementary facts.

(1) Both empirical traditions and theoretical traditions are traditions in their own right, with their own laws, objects, research procedures and associated philosophies.[20] Rationalism did not introduce order where before there was chaos and ignorance; it introduced a special kind of order, established by special procedures, and different from the order and the procedures of historical traditions.

(2) The theoretical approach had results in fields such as astronomy and mathematics. In the *Republic*, 530bff Plato advised the astronomers to construct abstract models and to 'disregard things in the heavens'. Those following the advice succeeded beyond expectations. But *the success could not be foreseen* and, besides, it did not immediately lead to *better numerical values* than, say, the Babylonian predictions which rested on different and more empirical principles.

[19] This applies also to Popper's account (*Conjectures and Refutations* (New York, 1962), 127; cf. 150 and passim) which has already been accepted by some scholars (see G. E. R. Lloyd, *Early Greek Science – Thales to Aristotle* (New York, 1970, 10ff). According to Popper, science (and rationalism in general) 'is differentiated from older myths not by being something distinct from myth, but by being accompanied by a second order tradition – that of critically discussing the myth . . . Before, there was only a first order tradition. A definite story was handed on. Now there was still, of course, a story to be handed on but with it went something like a silent accompanying text of a second order character: "I hand it on to you, but tell me what you think of it. Think it over. Perhaps you can give us a different story." This second order tradition was the criticial, or argumentative attitude.'

Popper's account is certainly better than more traditional accounts which merely observe a transition from one 'first order tradition' to another and more familiar first order tradition (such as materialism). But it overlooks that historical traditions have internal means of criticism and improvement and are therefore not in need of second order traditions and that *second order traditions often disturb historical traditions and rob them of their efficiency*.

[20] Scepticism, for example, became the house philosophy of empirical traditions in medicine.

(3) But the theoretical approach also ran into considerable difficulties both within itself, and in the attempt to conquer empirical traditions such as medicine. The *internal difficulties* are well known. They are: the paradoxes of Parmenides and Zeno; the discovery of irrational numbers; the semantic paradoxes; the difficulties of the theory of ideas; the sense–reason problem; the mind–body problem, and so on.[21] Some of these difficulties such as the problems of the continuum are still with us and they may be a reason for certain peculiar features of the quantum theory. They reveal a basic weakness of abstract thought, left to its own resources. Some philosophers and mathematicians have therefore suggested a partial return to more empirical traditions even right in the centre of mathematics (Brouwer, Wittgenstein, Polya and others).[22]

(4) The *external difficulties* of the abstract approach were soon noticed and criticized by the physicians of the Coan School in medicine.[23] Thus we read in ch. 15 of *Ancient Medicine*:[24]

> I am at a loss to understand how those who maintain the other view and abandon the old method in order to rest the *techne* upon a postulate, treat their patients on the lines of their postulate. For they have not discovered, I think, an absolute hot and cold, dry or moist, that participates in no other form. But I think that they have at their disposal the same foods and the same drinks as we all use, and to one they add the attribute of being hot, to another, cold, to another, dry, to another, moist, since it would be futile to order a patient to take something hot, as he would at once ask 'what hot thing?' So that they must either talk nonsense or have recourse to one of these known substances.

In this passage 'scientific' medicine is criticized for introducing abstractions such as an abstract theory of the elements without indicating (except by *ad hoc* translations which do not increase the ability to heal) how the abstractions are related to concrete things such as food, or particular afflictions of mind and body. These afflictions and the attitude that characterizes some abnormalities as illnesses, others as excellence, still others as temporary oddities (sudden rage, for example; cf. *A M*, 243, n.66 and text) are known to the patient and the physician not theoretically, on the basis of a 'postulate' but empirically, by participation in a historical tradition which the physician expands by developing special skills of recognition and healing. Not theory, but these skills are the essence of medicine. Or, to continue with the text[25]

[21] Some rationalists view a proliferation of internal problems as a sign of fruitfulness (ch. 10, n.25 and text) – a somewhat strange notion of fruitfulness.

[22] G. Polya still retains demonstrations as an essential, 'safe, final and uncontroversial' ingredient of mathematics; cf. *Induction and Analogy in Mathematics* (Princeton, 1954), v. This remaining dogmatism was removed by Lakatos; see *Proofs and Refutations* (Cambridge, 1978).

[23] For the division into schools see I. M. Lonie, 'Hippocrates and the Historians' *Hist. Sci.*, 16 (1978), nos. 31 and 32. Cf. also part 1, ch. 1 of H. L. Coulter, *Divided Legacy* (Washington, 1975).

[24] Ed. W. H. S. Jones (Loeb Classical Library, 1962) I, 41.

[25] Jones I, 53, my italics.

> Certain physicians and philosophers assert that nobody can know medicine who is ignorant of what man is; he who would treat patients properly must, they say, learn this. But the question they raise is one for philosophy [i.e. for abstract thought, not for the practice of healing]; it is the province of those who, like Empedokles, have written on natural science, what man is from the beginning, how he came into being at first, and from what elements he was originally constructed. But my view is, first, that all that philosophers and physicians have said or written on natural science *no more pertains to medicine than to painting.*

The issue between a scientific medicine and traditional forms of medicine such as American Indian medicine (often used by US physicians in the nineteenth century), herbalists, acupuncture etc. is a modern version of the problems alluded to in this quotation.[26] No satisfactory comparison between these systems exists. We know that scientific medicine has achieved cures in special domains that we cannot easily imagine repeated in other systems (though even here we must often rely on rumour rather than on fact; see my *SFS*, 88ff). The overall picture is far from clear, however. For example there is no evidence to show that the use of traditional non-scientific methods in medicine cannot terminate the stagnation of scientific cancer research.[27]

(5) Difficulties in ethics and politics were already noticed by Plato. Plato is one of the few philosophers who noticed the limits of a purely abstract approach and who both accentuated and tried to overcome them. The difference between the *Laws* and the *Republic* may be explained by the greater weight given to historical matters in the later dialogue.

(6) The history of the sciences provides us with an apparent paradox. The sciences and especially the natural sciences and mathematics seem to be theoretical subjects *kat'exochen*. They arose when Greek theoretical traditions replaced the empirical traditions of the Babylonians and the Egyptians (the invention of central perspective in fifteenth-century Italy is a similar development right in the centre of the so-called 'arts'). Now the interesting thing is that *after the first abstract steps these subjects became empirical (historical) traditions in their own right.* Abstract notions and procedures were used but in an intuitive way that often conflicted with their abstract definition (cf. ch. 5 on the older quantum theory). This has been realized only quite recently, after the breakdown of all attempts to give a 'rational' account of scientific change. Subjects can become complex and successful only by turning the abstractions they contain into concepts that are guided not by rules (except locally) but by the tact and the intuitions conferred

[26] Quite recently a congressional committee had to force cancer researchers to pay greater attention to the effects of nutrition.

[27] For the stagnation cf. the report by Daniel Greenberg 'The "War on Cancer" Official Fiction and Health Facts' *Science and Government Report*, IV (1 Dec. 1974).

upon its practitioners by a historical tradition: strictly speaking *all sciences are Geisteswissenschaften.*

4. ARISTOTLE

Aristotle had a clear insight into some important features of historical traditions, ancient myths included;[28] he regarded them as important parts of the history of man to be included in any complete account. This was the reason why he developed the history of ideas (ch. 9.9) and why he started every abstract discussion with a review of previous opinions, arguments, 'results'. Aristotle also accepted the new abstract ideas of the philosophers and he attempted a synthesis of historical and theoretical (abstract) thought. To achieve the synthesis he on the one hand related abstract ideas to commonsense by restricting their domain of application to mirror the fact that historical traditions do not judge absolutely, but adapt their judgements to circumstances. On the other hand he tried to base the restriction on principles and in this way replace tact and intuition by rules. This was a most delicate process of adaptation (*SFS*, 55ff) which not only made commonsense more rational, and reason more realistic, but which also gave rise to a great variety of new subjects: logic, rhetoric, psychology, metaphysics, physics, cosmology, ethics, politics, mathematical philosophy, theology, a theory of the arts; all of which have influenced Western thought down to the twentieth century.[29] Here was an early heroic time of rationalism when abstract reason not merely tried to *capture* the 'inherent rationality' of subjects that had arisen without it but *created* subjects and was therefore fruitful and not merely imitative.

Aristotle's general philosophy is a form of empiricism. Aristotle and modern empiricism both admit that our knowledge is often disfigured by error. They differ in their assumptions about the role that error is permitted to play (cf. *SFS*, 56ff). In Aristotle error distorts special perceptions but leaves the general features of perception untouched. Man commits sizeable mistakes, but it is always possible to reconstruct the general features of perception. This is another similarity with commonsense. Commonsense also admits that there are mistakes; it has found ways of eliminating or reducing them, but it will never assume that man awake can be completely misguided. Error is a local phenomenon; it does not distort our entire world view. Modern science, on the other hand (and the Platonic and atomistic philosophies it absorbed) assumes precisely such global distortions. When it arose in the sixteenth and seventeenth centuries it 'questioned an entire

[28] *Metaphysics*, 1074b1; cf. *On the heavens*, 270b9, *Meteorology*, 339b27, *Politics*, 1329b25; cf. also Werner Jaeger, *Aristotle* (Oxford, 1958), ch. 6.

[29] Think of the role of Aristotelianism in theological disputes down to our time and the quite different role it played in the history of dramatic theory (Corneille, Lessing, Brecht).

system and not only a detail; it attacked not only physics, but almost all sciences and all traditions . . .'[30]

In *AM* I have described and discussed some aspects of this process. In part 1, ch. 5 of *SFS* I added some further observations. I showed that most of the arguments that were used in astronomy were arguments in appearance only and that they were effective because astronomers had already changed their point of view. Like modern scientists the followers of Aristotle distinguished between basic theory and purely formal devices. They admitted that formal devices might show some internal harmony but added that harmony, taken by itself, it not a sufficient criterion of reality (e.g. the harmony of Schrödinger's formalism does not show that there exist 'objective' quantum waves, as Schrödinger had originally assumed). Yet this was precisely the position taken by some of the followers of Copernicus such as Rheticus, Maestlin, and Kepler (*SFS*, 51f): Aristotelian physics was rejected not because it was shown to be faulty, but because a tacit transition to a new criterion of reality gave force to considerations which otherwise could have been easily refuted.[31]

Now the Copernican revolution did not only contain changes in astronomy. Being based on a realistic interpretation of a new planetary theory, i.e. of new assumptions about the heavens and the earth, it also affected physics, cosmology, the theory of knowledge, theology, the calculation of tables and general philosophy (e.g. nature of man). The usual attempts to give a 'rational explanation' of it only deal with the theory of planetary motions and even here it is assumed that everybody went the same way: all good astronomers had exactly the same attitude towards Copernicus and this attitude was based on certain epistemological standards. The assumption is not correct.[32] Outstanding astronomers such as Tycho Brahe turned against Copernicus, others accepted some technical details without accepting the philosophical turn, still others defended him wholeheartedly, but for different reasons. It is this mixture of very different 'microprocesses' which *in astronomy* led to the macroprocess: acceptance of Copernicanism. Neither the macroprocess nor the microprocesses agreed with accepted standards of rational behaviour (and they don't agree with modern standards either; cf.

[30] Aristotle, *Physics*, 253a31ff (on Parmenides).

[31] One might be inclined to regard Galileo's dynamical investigations as a scientific refutation of Aristotle's physics. Now while Galileo (and those of his predecessors on whom he builds) have certainly refuted *special hypotheses* of the Aristotelian opus (such as the theory of antiperistasis) they did not refute the *basic laws* such as the law of inertia; for this law could always be combined with the impetus theory and so made to agree with the facts. The manner of agreement was not *ad hoc* for it rested on generalizations from other fields (behaviour of heat etc.). The impetus theory, however, was incompatible with Galileo's new *principle of relativity* which in turn was introduced to 'save' the Copernican hypothesis; cf. *AM*, ch. 8.

[32] Some of the 'microprocesses' which refute the assumption were examined by Professor Westman.

the list in *SFS*, 45f). Adding the remaining fields further complicates the situation. It needs an almost Hegelian faith in the power of reason to assume that this entire mass of theories, boundary conditions, assumptions of faith, presuppositions all moved side by side, in accordance with simple and easily identifiable standards. Being part of this entire complex process the attitude towards the validity and the alleged refutation of Aristotle's doctrines becomes a very subtle and difficult problem. To start with, Aristotle survived much longer and with much better reasons than the more superficial historians have led us to believe. Aristotle's physics survived until late into the eighteenth century not because of its sly adaptive procedures[33] but because the ideas were still extremely useful in research.[34] Harvey[35] retained and employed Aristotelian principles in his studies and the functional approach which is characteristic of Aristotle has remained important. Aristotle's law of inertia underlies all work in bacteriology and virology; most of the discoveries made in these fields would have been impossible without it (the same applies to important biological research dealing with the question of spontaneous generation). Copernicus adapted his ideas to the Aristotelian philosophy arguing that they were required rather than forbidden by this philosophy (*AM*, 93). In astronomy Aristotelian views disappeared for a variety of reasons, only some of them 'rational'. Outside astronomy there arose a fashion to malign Aristotle. The fashion (which was not widespread) had a great variety of causes: the opposition between the schools and 'practical people'; the connection of Aristotle with apparently backward social forces; the growth of new philosophical parties such as the Cartesians; the fact that many Aristotelians were mediocre textbook rationalists (one should compare the situation at the LSE after the death of Imre Lakatos); a lack of familiarity with the basic principles of the Aristotelian philosophy so that slogans took the place of arguments; the rise of a new specialist ideology that favoured mathematics and looked with boredom (or contempt) at fundamental research (a little of this seems to be present already in Maestlin, Kepler's teacher); the wish for 'content increasing' research i.e. the wish to penetrate beyond appearances to a world that was hidden from the common folk but accessible to mathematicians and philosophers; and so on. Today, after the arrival of the theory of relativity, quantum mechanics, the thermodynamics of open systems (Prigogine) and the most recent developments of the science of mechanics itself (Moser) it has become evident that Aristotelian physics with its emphasis on well-structured processes with a beginning, a middle, and an end, and its denial of an absolute void provides a much more adequate natural

[33] As is hinted by Edward Grant, 'The Longevity of Aristotelianism', *Hist. Sci.*, 16, (1978) 94ff.

[34] Cf. the account of electrical research in John Heilbronn's marvellous *Electricity in the 17th and 18th Centuries* (Berkeley and Los Angeles, 1979), e.g. 101ff.

[35] Cf. Walter Pagel, *William Harvey's Biological Ideas* (Basle, 1967).

philosophy than the mechanical point of view of the seventeenth and eighteenth centuries which has retained its influence up to the present day. And those who still rant and rave against him turn out to be 'crude animals who bark at things they do not comprehend' (Albertus Magnus[36]).

We have seen that Aristotelianism contained not only specialist assumptions such as the idea of the central symmetrical universe; it also contained a general theory of man and of his relation to the world. According to this theory man has ideas as well as experiences. His ideas may be very abstract and removed from experience, but according to Aristotle their content is still determined by experience: *it is commonsense and not the ideology of intellectuals that determines whether or not something exists and, if it exists, what properties it has.* Intellectuals may clarify the various bases of knowledge and the principles that underlie research; they may show us what we can learn from these bases and these principles but their ideas occasionally take flight and get lost in abstractions. Such abstractions are not always useless; they may help us to arrive at useful *predictions*. The *nature* of things, however, depends on the philosophical commonsense that is the result of Aristotle's synthesis of philosophy and common commonsense (see above). The instrumentalism of Aristotle is therefore a most interesting *anthropological* (*and ethical*) *thesis*. It is surprising to see how a philosophy that asserted the common ground of mankind and wanted to build knowledge on it became a school philosophy, separated from science as well as from commonsense, and how it was finally attacked by the very same 'common' people whose ideas it had tried to explore: even the best philosophy degenerates when it gets into the wrong hands.

5. PHILOSOPHICAL STANDARDS AND PRACTICAL METHODS

Among the late medieval and early modern opponents of Aristotle we find both philosophers and practical groups. It is interesting to compare the ideas of the two groups.

The fourteenth century saw the rise of artisans, artists, and sailors to importance and respectability.[37] Navigators discovered the West African coast, found the best routes to the East,[38] increased the power of Spanish and Portuguese kings, corrected maps, and refuted ancient geographical ideas; artists found the laws of central perspective and corrected them to

[36] Comment on the letters of Dionysius Areopagita 7, 2B *Opera Omnia* (Borgnet) 14, 910.

[37] For the artists cf. Arnold Hauser, *The Social History of Art* (New York), II as well as chs. 2 and 3 of R. & M. Wittkower, *Born under Saturn* (New York, 1963). For Artisans cf. P. Rossi, *Philosophy Technology and the Arts in the Early Modern Era* (New York, 1970).

[38] A lively account whose occasional mistakes are more than balanced by the splendid story-telling power of the author is Stefan Zweig, *Magellan* (Frankfurt, 1977). Pre-Columbian voyages of discovery are discussed with ample documentation in R. Henning, *Terrae Invognitae* (Leiden, 1938), IV.

establish a closer fit between geometry and human vision;[39] artisans contributed to the knowledge of metals and minerals; herbalists enriched medicine.[40] The eyeglass was known as early as the thirteenth century, and the telescope was invented in Holland by artisans long before there was a 'scientific' understanding of its principles.[41] These inventions, their consequences, and their importance for knowledge occurred and were discussed outside the schools, almost without help from contemporary scholarship.[42] The successes achieved led to the belief that truth could be obtained without the help of the traditional guardians of knowledge and without paying attention to the standards of the schools. 'Through practice' writes Palissy[43] 'I prove that the theories of many philosophers, even the most ancient and famous ones, are erroneous in many points. Anyone can ascertain this for himself, in two hours, merely by taking the trouble to visit my workshop.' 'But I doe verily thinke' writes the sailor Robert Norman whom Gilbert mentions in his writings

> that notwithstanding the learned in the sciences, being in their studies amongst their books, can imagine great matters, and set downe their farre fetched conceits in faire flowe, and with plentiful wordes, wishing that all Mechanitians were such, as for want of utterance, should be forced to deliver unto them their knowledge and their conceits, that they might flourish upon them, and applye them at their pleasures; yet there are in this land diverse Mechanitians, that in their severall faculties and professions have the use of those arts at their fingers ends. And can applye them to their severall purposes, as affectionately and more readily than those who would most condemn them.

And Agricola demands for mining the dignity of a philosophical enterprise.

The standards, procedures, ideas of these discoverers, inventors, thinkers are partly intuitive – they are not explicitly described and one has to read them off their actions – partly determined by their professions. Occasionally one finds explicit accounts which are criticized, developed and turned into fully fledged *philosophies of research*. Such philosophies do not exclude authority; quite the contrary: old authorities, Aristotle included,

[39] Cf. E. Panofsky, 'Die Perspektive als Symbolische Form' reprinted in *Aufsaetze zu Grundfragen der Kunstwissenschaft* (Berlin, 1974). This essay has influenced all subsequent discussions on the history of perspective. It is interesting to see that artists soon discovered the difference between mathematical construction and the laws of vision. Mannerism was in part a consequence of this discovery (cf. Arnold Hauser, *Der Manierismus* (Munich, 1964)). 'Scientific' optics disregarded the conflict and gave a wildly incorrect account of the laws of vision down to the twentieth century: *A M*, ch. 10, esp. n.55. Cf. also V. Ronchi, *Optics, the Science of Vision* (New York, 1957).

[40] Cf. the work of the Paracelsians as well as L. Thorndyke, *History of Magic and the Experimental Sciences* (New York, 1957), II and VI.

[41] V. Ronchi, *Histoire de la Lumière* (Paris, 1956).

[42] Cf. the older account in L. Olschki, *Geschichte der Neusprachlichen Wissenschaftlichen Literatur* (Reprint Vaduz, 1965), I.

[43] Quoted from ch. 1 of Rossi, *Philosophy, Technology and the Arts.*

are greedily studied, for one regards them as being superior by far to the school philosophies of the time. Lorenzo Ghiberti tries hard to learn, at an advanced age, the history of the arts so that he can write his *Commentarii*[44] and the influence of the new scholars who arrived after the fall of Constantinople is soon noticed in all subjects.[45] Thus the authorities are studied very carefully, but they do not have the last word.

The reason is that their judgement is balanced by an appeal to *experience*. This second and most important source of knowledge is neither the experience of the Aristotelians which does not comprise professional expertise; nor is it the sense-data experience of the sceptics and of later philosophers that has been cleaned of all prejudice. It is rather the changing ability of the professional to deal with his surroundings; it uses the *schooled eye*, the *practised hand* of the artisan, the navigator, the artist and it develops with his craft. The relation between authority, experience and knowledge, however, is seen in the following way and guided by the following rules.

(i) Read what others have said about the things you are interested in but don't trust them too much. Respect your predecessors, but don't be their slave!

(ii) Give due regard to the experience of your profession, i.e. use both the information accumulated by it and the observational skills you possess and try to decide doubtful points with their help. However, do not rest content with this information and these skills, but try to improve them!

(iii) Make assertions which are plausible in the light of (i) and (ii) for they alone contain a knowledge of things.

Note the loose and informal character of these rules (which are not found in books, or tracts but can be abstracted from practice and popular sayings). Thus the third rule does not assume a logic of induction or any other formal or formalizable relation between 'knowledge' and 'experience'. It does assume that there are people who have learned to master a certain historical tradition, who know how to make plausible assumptions in concrete circumstances, and it advises them to use this ability to construct knowledge. Rules of this kind reflect the new *historical traditions* which are now being produced and which have numerous discoveries to their credit. They even fit the much more sophisticated procedures of later

[44] Cf. the somewhat unjust report of Olschki, *Geschichte* i, with Krautheimer's 'The Beginnings of Art Historical Writing in Italy' reprinted in R. Krautheimer, *Studies in Early Christian, Mediaeval and Renaissance Art* (New York, 1969). Cf. also R. Krautheimer & Trude Krautheimer-Hess, *Lorenzo Ghiberti* (Princeton, 1956).

[45] S. Y. Edgerton Jr, *The Renaissance Rediscovery of Linear Perspective* (New York, 1975), ch. 8 conjectures that Ptolemy's *Geographia* was a connecting link between the architecture of Brunelleschi and his investigations concerning linear perspective. The latter did not develop out of architectural drawing and it did not bring it about; both together were inspired by the attitude which arose in Florence under the influence of the new world atlas. Krautheimer & Krautheimer-Hess, *Lorenzo Ghiberti*, ch. 16 assume that the linear perspective of Brunellesci developed out of his architectural designs.

centuries (cf. ch. 5, on the difference between scientific and philosophical procedures). Science has always been a matter of context-dependent plausibility and not of a context-independent 'organon of thought'.[46]

This practical and highly successful philosophy of research[47] was soon modified and almost entirely replaced by *theoretical traditions* and corresponding methodologies. The resulting 'theories of scientific method' or 'theories of scientific rationality' have accompanied science up to the present day, they have beclouded our understanding of science and have occasionally interfered with the business of science itself. The following rules are examples of the ideas we encounter (but there are many other rules, held together by different ideologies):

(i′) eliminate prejudice,
(ii′) pay attention to experience, and
(iii′) make your ideas consistent with experience or derive them from experience.

The rules are very similar to the practical rules mentioned earlier. They can therefore live on their popularity. Every triumph of the historical traditions to which the earlier rules belong is also a triumph for the theoretical traditions that form the background of the new rules. But the difference in content is considerable! While rule (i) advises us not to take authorities too seriously, rule (i′) demands that they be removed. While rule (ii) makes use of the skills that an individual has learned and of the information that is at his disposal (which means that an entire tradition is mobilized at every single step) the 'experience' or rule (ii′) is a philosophical fantasy that occurs nowhere in nature and society and has to be manufactured by eliminating all skills, all information, all previous knowledge. While rule (iii) encourages the researcher to adapt his conjectures to a background of ideas, assumptions, even unconscious reactions where the kind of adaptation required depends on circumstances as mastered by the 'tact' inherent in a certain tradition, rule (iii′) introduces some abstract

[46] 'He seems to be very faulty' writes Descartes with the despair characteristic of the systematic philosopher confronted with the achievements of a scientist who works in this fashion (letter to Mersenne of 11 October 1638; quoted from Stillman Drake, *Galileo at Work* (Chicago, 1978), 387f; cf. also my account of Galileo's method in ch. 11 of *Der Wissenschaftstheoretische Realismus und die Autoritaet der Wissenschaften* (Wiesbaden-Braunschweig, 1978); for a comparable modern reaction cf. my remarks on Lakatos in vol. 1, ch. 16, n.102) on continually making digressions and never stopping to explain completely any matter [Galileo's ideas are adapted to concrete research problems many of whose aspects seem accidental to 'the systematic epistemologist'] – which shows that he has not examined things in order [the order of scientific research is not the order of philosophical explanation; cf. again ch. 5] and that without having considered first causes [i.e. without having paid attention to the basic notions of a particular philosophical system] he has only reasons for some particular effects and thus he has built without a foundation.'

[47] Aristotle himself never applied the 'methods' one constantly ascribes to him. *Epagoge* is not induction in the modern sense but rather a psychological process that makes us discover the general features underlying a particular effect.

relations, a 'logic' and is ready to accept only ideas that can be linked to the abstract evidence of rule (ii') in the manner stipulated by these relations. No doubt rules (i') to (iii') were influenced (e.g. in England; see *A M*, 46, n.13) by certain tendencies in *theology* that wanted to replace traditions by the pure and unadulterated word of God. Indeed, the new idea of an unprejudiced *experience* that has been cleaned of all human opinion has much in common with the new idea of an unprejudiced *divine decree* that affects the faithful directly, and without mediation through papal declarations, church councils, philosophical speculation. And like this idea it is refuted by the very simple arguments which Father Veron used against the French Protestants:[48] the method is not just impractical, it is an impossibility.

From now on the development is as follows. Rules (i') to (iii') and the underlying philosophy are criticized *abstractly*, i.e. one tries to find new rules and standards that are no longer exposed to Verons (or Hume's) objections but agree with the laws of logic. Nobody asks if the new rules are also useful for doing research (there are some exceptions to this statement, but they are not very numerous). The theory of knowledge becomes more and more removed from scientific practice while the increasing technicality of its procedures and the quasi-scientific terminology create the impression of progress and sophistication. The development begins with Newton's philosophy of science as described in the next chapter.

Newton uses his philosophy to protect a point of view in physics (his theory of light) that is beset by empirical and conceptual difficulties. He adapts the presentation of the *Principia* to it thus creating the impression, still strong towards the end of the nineteenth century, that the theory of gravitation can be derived from an experience of the kind described in (ii'). Even twentieth-century philosophers such as Ernest Nagel in his *Structure of Science* (reviewed in ch. 3) and twentieth-century scientists such as Max Born[49] still accept Newton's form of empiricism.

In the nineteenth century Newtonianism was criticized by philosophers and scientists. *John Stuart Mill* introduced proliferation (proscribed by Newton) in his *On Liberty* (see ch. 4) but without any effect on the philosophy of the natural sciences. *Hegel* developed a theory of conceptual change that undercut all existing forms of empiricism and he also criticized Newton's famous 'derivation' of the laws of gravitation from the facts.[50] His

[48] The argument which is described in R. H. Popkin, *The History of Skepticism from Erasmus to Descartes* (New York, 1964) anticipates Hume and the similar arguments of twentieth-century philosophers of science. It points out that a basis in the sense of (i') cannot be found without using the 'prejudices' (i') wants to eliminate; that, if found, it could not be understood without such prejudices; and that it would not be possible to derive consequences.

[49] *Natural Philosophy of Cause and Chance* (Oxford, 1958).

[50] *Encyclopädie der Philosophischen Wissenschaften*, ed. G. Lasson (Leipzig, 1920), 236, lines 11ff and 237, lines 21ff (perturbations). For Hegel, Engels, Lenin and Bohm, see ch. 4.

ideas influenced dialectical materialists but had little effect elsewhere. Both Mill and Hegel subjected standards of rationality and rules of method to the ongoing process of research and restored the point of view that underlies (i)–(iii) above. Their methodology was practical and research immanent, not philosophical and research transcendent, the traditions to which they appealed historical (empirical), not abstract. Scientists also tried to free themselves from the restrictions of a philosophical method. Maxwell, Helmholtz, Hertz, Boltzmann, Mach, Duhem all favoured a methodological pluralism guided by the examples of past research over research-transcendent standards. Each of these scientists was of course in favour of certain procedures and against others but they all agreed that such *personal preferences* must not be turned into *'objective' principles*. 'The best means of promoting the development of sciences' writes *Pierre Duhem* after a vigorous diatribe against model building[51] 'is to permit each form of intellect to develop itself by following its own laws and realising fully its type'. 'I must admit' writes *von Helmholtz*[52] 'that so far I have retained the latter procedure [mathematical equations instead of models] and felt safe with it – but I would not like to raise general objections against a way which such excellent physicists . . . have chosen.' *Boltzmann*, after a survey of new methods in theoretical physics objects to regarding any one of them, old or new, as the only acceptable one.[53] The pluralism of methods corresponds to the pluralism of ideas about physical reality, which I have already mentioned (vol. 1, ch. 1, sections 2 and 3). The reader should also notice that these considerations cut across the distinction between a context of discovery and a context of justification which philosophers have introduced to protect their prescriptions from a clash with scientific practice: some scientists refuse to contradict low level laws, others want to retain high level theories and make them measures of all science, still others hope that new discoveries will arise from a conflict with laws and principles, principles of logic included. Boltzmann and Mach treat the history of science as a special instance of the history of species and regard the ideas of their time as late but by no means final results. In chs. 5 and 6 I have given a more detailed account of *Mach*'s philosophy of science, its relation to Einstein's work as well as of the deterioration that set in during the twentieth century.

For the splendid proliferation of ideas comes to an end when the philosophy of science is finally turned into a special subject with special standards and an organon of its own: formal logic. As always maturity in a narrow domain means illiteracy elsewhere. A boring and barren theoretical tradition replaces the exciting debates of the nineteenth century, attention to topics without relevance to science replaces the many fruitful suggestions

[51] *The Aim and Structure of Physical Theory* (New York, 1962), 99.
[52] Introduction to Heinrich Hertz, *Die Prinzipien der Mechanik* (Leipzig, 1894), xxif.
[53] *Populäre Schriften* (Leipzig, 1906), ch. 1, esp. p. 10.

that came out of these debates. The old issue between historical traditions and abstract traditions opens up again, *but it is now science that is being distorted, with philosophy of science as the distorting abstract tradition* (cf. (6) of section 3 of this chapter). Chapter 5 gives a general account of this development with Mach as a starting point. Popper's 'philosophy' is an excellent specific example.

6. POPPER, KUHN, LAKATOS AND THE END OF RATIONALISM

Popper eclectically combined (1) the pluralism of Mill and of late nineteenth-century science, (2) Mill's account of the hypothetico-deductive method and his emphasis on negative arguments (ch. 9.13, n.50) and (3) the objections which some scientists raised against *ad hoc* hypotheses which are just repetitions of earlier objections against occult qualities. The non-technical version of the resulting 'philosophy', which Popper describes in *Conjectures and Refutations* (New York, 1962) (cf. also ch. 11.2) and which makes *criticism* an important part of science, contains valuable though hardly original observations on the origin and growth of rational knowledge. Though exclusive in rhetoric this version is not exclusive in practice: even dogmatic ideas can be developed (and often have been developed) from a criticism of more openminded philosophies. 'Anything goes' is an obvious practical consequence of such a 'critical rationalism' which Popper used to introduce by saying that although he was a professor of scientific method he could not act accordingly, for 'there is no scientific method'. Ernst Mach would have been very pleased with this statement (cf. ch. 6, nn.50–2).

But Popper and his pupils also developed and now defend a more technical version. By now this technical version has become a veritable malaise. The aim is no longer to understand and, perhaps, to aid the scientists; nor is there any attempt to check the version by a comparison with scientific practice. The aim is to develop a special point of view, to bring this point of view into logically acceptable form (which involves a considerable amount of rather pointless technicalities) and then to discuss everything in its terms (cf. Mill's observations on such developments as quoted and discussed in vol. 1, ch. 8). Not the ever-changing demands of scientific research but the rigid requirements of an abstract rationalism decide about the form and the content of the principles accepted.

These drawbacks are a direct consequence of the underlying philosophical aim. 'Epistemology must provide a strict and general criterion that enables us to separate statements of empirical science from metaphysical statements' writes Popper in his *Grundprobleme*[54] which preceded the *Logic of Scientific Discovery*. He takes it for granted (*a*) that the suggested conceptual

[54] *Die Beiden Grundprobleme der Erkenntnistheorie* (Tübingen, 1979), 422.

dichotomy (between empirical and metaphysical statements) corresponds to a real separation between statements that are part of scientific traditions, and (*b*) that traditions which do not know the separation or contain it only in a blurred form are improved by eliminating the blur. More recent historical research has shown that (*a*) is definitely mistaken: there does not anywhere exist a cluster of statements that is scientific in Popper's sense and sufficient for bringing about the results for which the sciences are famous. (*b*), on the other hand, has hardly ever been examined. And yet a closer look at the growth of rationalism in Greece, at the comparative effectiveness of 'scientific' medicine and various folk medicines, at the comparative effectiveness, in special situations, of oracles and a 'rational discussion' (described vividly by Evans-Pritchard) shows that this assumption, too is highly questionable. It is not too far from the truth if we say, with Wittgenstein, that the epistemological approach implicit in the above quotation erects castles in the air which have but little to do with things we regard as beneficial and important. A closer look at the technical version fully confirms this diagnosis.

According to the technical version, science proceeds by identifying problems and solving them with the help of hypotheses which are(*a*) relevant, (*b*) falsifiable and (*c*) richer in content than the descriptions from which the problems arose. Having found a suitable hypothesis one (*d*) tries to falsify it and opposes any attempt to explain away difficulties. Falsification leads to the new problem of why the older theory succeeded where it did and why it failed. This problem must again be solved in accordance with (*a*), (*b*), (*c*) and (*d*), i.e. by hypotheses which are richer than both problems and falsifiable. And so science advances by conjectures and refutations from local regularities to comprehensive conceptual schemes. There is no guarantee that we shall always be able to solve the problems we face but if we do, in accordance with (*a*), (*b*), (*c*), (*d*), then progress is guaranteed and we also know wherein it consists.

Now we may regard (*a*) to (*d*) as *useful hints* for the scientist which he may adopt but which he may also overrule if his problem situation demands it; on the other hand we may regard them as *necessary conditions* of a rational approach and, therefore, as invariable features of all important scientific work. Popper (cf. his objections to 'naturalism') has often interpreted the conditions in the second way; Popperians have never considered any other interpretation.[55] The interpretation is clearly unsatisfactory for a number of reasons.

The *first reason* is that theory exchange is not always by falsification. Examples are the Copernican revolution and the special theory of relativity. *There is no refuting fact*, or set of facts, that can explain the removal of Ptolemy, Aristotle, or the literal interpretation of the Bible and there is no

[55] An apparent exception is Lakatos, but he is not a Popperian; cf. ch. 9, n.57.

refuting fact that can explain the removal of the Lorentz theory of electrons and that was interpreted by the participants as demanding the removal. Michelson's experiment which is often mentioned in connection with the latter case was explained by Lorentz, and in a 'content increasing way' as Popperians are fond of saying (ch. 10.6). We can of course devise an interpretation in which the experiment refutes any ether theory (assuming we choose suitable boundary conditions – not at all an easy matter!) *but only after the development that the refutation is supposed to cause has taken place*. Similar remarks apply to whatever 'facts' one might want to mobilize against the Ptolemaic–Aristotelian point of view (*SFS*, 49ff).

The *second reason* is that the 'meaning' of a hypothesis often becomes clear only after the process that led to its elimination has been completed (cf. ch. 10.14). White ravens refute 'all ravens are black'. But a raven that has been painted white, or has fallen into a bag of flour, or has been bleached by industrial fumes does not count as a white raven (conversely, a bird that has been blackened by analogous processes does not count as black). Does a raven whose metabolic processes make him white or whose genetic makeup has been interfered with so as to make him white count as a white raven? Matters such as these are decided by studies in colour variation, i.e. *after* many potentially refuting facts have been considered: the content of the theory we want to test and our decisions about falsifying instances are not as independent as a strict theory of falsification would want them to be. Combining the demand to fix this content before tests and independently of them with the assertion that any such fixing belongs to the context of discovery and that the context of discovery may last a considerable time does not solve the conundrum, for now we must admit that theories are often abandoned long before their context of discovery has come to an end.

Third, there are many cases where a transition to a new theory involves a change of universal principles (*AM*, 269) and so breaks the logical links between the theory and the content of its predecessor. This does not worry scientists who have many ways of choosing between 'incommensurable' points of view (vol. 1, ch. 1, n.37), but it conflicts with the technical version (verisimilitude; content increase).

Fourth, contents do not always increase; they occasionally shrink, or are adapted in an *ad hoc* manner. Examples are the quantum theory which *assumes* but does not *explain* classical states of affairs and certain approximation methods in the general theory of relativity (*AM*, 63). The rise of scientific psychology was accompanied by a considerable reduction of content; so was the elimination of the ether theories. Some Popperians object that these examples deal with theoretical statements but this is to withdraw to a naive observationalist philosophy (*SFS*, 216).

Fifth, *ad hoc* adaptations are often the right step to take. Thus some researchers assumed, in the early history of electricity, that 'hair, leaves, twigs and other chaff drawn by amber contained a common buried principle'.[56] Gilbert, being a good Popperian, 'thought the idea too silly to require refutation'. Yet the idea was a step in the right direction. Similar remarks apply to some of Galileo's moves (*A M*, 93ff) and to many other episodes in the history of science.

Sixth, the demand to look for refutations and to take them seriously leads to an orderly development only in a world in which refuting instances are rare and turn up at large intervals, like large earthquakes. In such a world we can build, improve, live peacefully with our theories from one refutation to the next. But all this is impossible if theories are surrounded by an 'ocean of anomalies' (vol. 1, ch. 6.1), unless we modify the stern rules of falsification, i.e. unless we use them as rules of thumb, or as temporary ingredients of rationality, not as necessary conditions of scientific procedure.

Seventh, the demand for unceasing increase makes sense only in a world that is infinite both quantitatively and qualitatively. In a finite world, containing a finite number of basic qualities, or 'elements', the aim is first to find these elements and then to show how novel facts can be reduced to them with the help of *ad hoc* hypotheses. Genuine novelty counts as an argument against the methods that produce it.

Eighth, content increase and the realistic interpretation of the ideas that bring it about may be rejected for *ethical* or *political* reasons (this point was briefly made in the introduction (to vols. 1 and 2). For example, we may want to view humans in a 'subjective' fashion, using as well as attributing affection, concern, pity, and 'deeper' properties, belonging to a soul. The demand for content increase together with the demand to turn the most general theories in a certain domain into measures of reality often conflicts with this wish. Deciding to uphold the subjective view we also decide against realism and content increase. This, to my mind, is the most powerful argument for regarding (*a*), (*b*) and (*c*) above as rules of thumb rather than as necessary conditions for science and of knowledge in general.

Kuhn has asserted that this is how science actually proceeds. According to Kuhn science is a *historical tradition* in the sense explained earlier; it is not subjected to external rules, the rules that guide the scientist are not always known, and they change from one period to the next. Understanding a period of science is similar to understanding a stylistic period in the history of the arts. There is an obvious unity, but it cannot be summarized in a few simple rules and the rules that guide it must be found by detailed historical studies (the philosophical background is explained by Wittgenstein; see ch. 7). The *general* notion of such a unity, or 'paradigm' will therefore be poor and it will state a problem rather than providing a solution: the problem of filling an elastic but ill defined framework with an ever-changing concrete historical content. It will also be imprecise. Unlike the sections of a theor-

[56] Heilbronn, *Electricity in the 17th and 18th Centuries*, 176f.

etical tradition which all share certain basic concepts the sections of historical traditions are connected only by vague similarities. Philosophers interested in general accounts and yet demanding precision and lack of ambiguity (as Laudan does cf. ch. 11.2) are therefore on the wrong track; there are no general and precise statements about paradigms. Polanyi and Kuhn make the surprising assertion that even the most abstract sciences are historical traditions in this sense.

Lakatos is the only modern philosopher of science in the Anglo-American tradition who has interpreted the problem of rationalism as a historical problem and who has tried to solve it historically by showing that all scientific developments after the Copernican revolution happen to have certain abstract features in common: science is a theoretical tradition even though the abstractions it contains are already very thin and evanescent. How thin is explained in ch. 10. In trying to establish his thesis Lakatos has unearthed some very interesting features of scientific change and he has come closer to science than any philosopher of science before him (in the twentieth century, that is). But he has not succeeded in showing that the material he examines has indeed the underlying 'abstract' structure. His progress is in partial historical insight, not in complete philosophical penetration. (The same applies to *Laudan* who, despite his belligerent assertion of novelty and improvement, has copied Kuhn and Lakatos in every detail.) These two writers have created the impression of possessing a universally valid *theory* of science when in fact they only have loose *terminology* without any general historical regularity to give it substance. This, then, is the end of the twentieth century dream of a scientific rationalism. Once, long ago, the belief in general laws of reason led to marvellous discoveries and so a tremendous increase in knowledge. Early physics, astronomy, mathematics were inspired by this belief as was the magnificent Aristotelian opus. In those times, and even more recently, during the rise of modern science and its twentieth century revisions, Lady Reason was a beautiful, helpful though occasionally somewhat overbearing, goddess of research. Today her philosophical suitors (or, should I rather say, pimps?) have turned her into a 'mature', i.e. garrulous but toothless old woman.

7. POLITICAL CONSEQUENCES

The decline of rationalism has interesting and troublesome political consequences.

It is generally agreed that a free society must not be left at the mercy of the institutions it contains; it must be able to supervise and to control them. The citizens and the democratic councils that exercise the control must evaluate the achievements and the effects of the most powerful institutions.

For example, they must evaluate the effects of science and take steps (withdrawal of financial support; reduction of scientific influence in elementary and highschool education; limitation and perhaps complete removal of academic freedom; and so on) if these effects turn out to be useless or harmful. To evaluate, the citizens need intellectual guides, they need standards. Now if standards for the evaluation of scientific research are research immanent, if they change as research proceeds and if their change can be controlled and understood only by those immersed in research, then a citizen who wants to judge science must either become a scientist himself, or he must defer to the advice of experts. A democratic control of science (and of other institutions) is then impossible.

This is indeed the conclusion that has been drawn by Michael *Polanyi*, one of the few twentieth-century scientist-philosophers to notice and assert the research immanence of scientific standards. According to Polanyi there is no way in which outsiders can judge science. *Science knows best.* (Kuhn and Holton give the same answer, though in more muted terms.) Are we to concede that the end of the philosophy of science also means the end of a democratic control of science and of scientists?

According to *Lakatos* who has devoted much attention to this problem a democratic control of science (and of other institutions) is possible only if we have criteria, or standards, which are *respectable* and which can be *separated* from scientific practice. The standards must be respectable, for we want to make a serious choice and not merely follow the whim of the moment. And they must be separable from scientific practice, for outsiders such as common citizens must be able to learn, use and apply them without becoming scientists. Lakatos' interest in general and situation-independent standards has philosophical as well as political motives.

To guarantee the respectability of his standards Lakatos connects them with science. To guarantee their separability he connects them only with special parts of science which can be expected to have certain features in common. The parts Lakatos chooses are achievements which everybody regards as important and path breaking. Newton's mechanics, Darwin's theory, the rise of the special and the general theory of relativity are achievements of this kind. Lakatos admits that science as a whole can be grasped by research-immanent standards only. The standards which judge science at its best, however, can be detached from scientific practice and understood independently of any mastery of it. Moreover, they have bite, for they can be turned against scientific developments that do not compare favourably with the events from which they were abstracted (according to Lakatos, modern elementary particle physics, empirical sociology, psychoanalysis, astrology, parapsychology are such developments). Lakatos advises foundations, political bodies, individual citizens to use the standards in their evaluation of the sciences and to withdraw money, political support, educational authority, etc. from developments that do not conform to them.

We must admit that Lakatos has recognized a most important problem, but he has not solved it. The standards he proposes do not fit the parts of science he has chosen as a basis (this is shown in ch. 10), and we never hear why a deviation from the standards should be regarded as a disadvantage and not as an improvement. Standards changed from Aristotle to modern science; everybody now regards this as an improvement. Assume they change again in the twentieth century, for example in elementary particle physics. Why is this change not a further improvement? We get no answer. Thirdly, the standards permit us only to compare one part of science with another; they do not help us to judge science as a whole: Lakatos shares Polanyi's view of the role of science in society except that he chooses part of science as his measure while Polanyi chooses all of science. Hence, if Polanyi is a 'Stalinist' (or 'elitist'[57]) then so is Lakatos, except that he bases his Stalinism on a different and more narrow basis than Polanyi. The second point (imperviousness of the standards to a change of science) also shows that the sciences enter the argument as windowdressing only. They are used for their propagandistic value as long as they agree with a preconceived philosophical idea (the idea of content increase); they are dropped as soon as they move away from the idea. Lakatos' elitism is therefore the elitism of a *philosophical clique* that wants to intimidate people in exactly the same manner in which science has intimidated them so far.

With this we are back to our original problem: how is a citizen going to judge the suggestions issuing from the institutions that surround him and how is he going to judge these institutions themselves? He needs criteria and standards; this is what our intellectuals tell us. What standards is he going to use?

The reply to this question is obvious. Scientific standards, we found, are not imposed upon science from the outside but are subjected to the practice of research just as scientific theories are subjected to the practice of research. In part 1 of *SFS* I showed that the same is true of all traditions. Each tradition, each form of life has its own standards of judging human behaviour and these standards change in accordance with the problems that the tradition is constrained to solve. Rationalism is not a boundary condition for traditions; it is itself a tradition, and not always a successful one. There exists therefore a plurality of standards just as there is a plurality of individuals. In a free society, however, a *citizen will use the standards of the tradition to which (s)he belongs*: Hopi standards, if he is a Hopi; fundamentalist Protestant standards, if he is a fundamentalist; ancient

[57] For 'elitists' cf. Lakatos *Mathematics, Science and Epistemology: Philosophical Papers* II, ed. J. Worrall & G. Currie (Cambridge, 1978), 114f. In talks and private conversations Lakatos used 'Stalinist' instead. For the faults of Lakatos' philosophy of science cf. ch. 10.

Jewish standards if he belongs to a group of people trying to revive ancient Jewish traditions; nor must we forget special groups which, realizing that they have special interests and ideas try to act in accordance with them – I am thinking of the women's movements, gay liberation, oecological groups and so on. Of course, all the groups need knowledge to apply the standards they use but the epistemic criteria which decide what is knowledge and what is not are determined by the traditions themselves and not by outside agencies. It is also clear that people learn and adopt ideas from other traditions, but this process again depends on the standards of the tradition that does the adopting. Finally, we must not overlook the fact that today almost all traditions are parts of larger units – they are parts of a city, a cluster of cities, a state, a confederation of states – and that they are constrained by the institutions and the laws of these units. It again depends on the traditions how they will deal with such constraints, for example, how they will use them to further their own interests. Some citizens of the state of California have used state laws to introduce ideas of *Genesis* into biology textbooks and to remove passages which presented evolution as a fact. Black muslims became capitalists to increase their fiscal and spiritual independence. The citizens of Puerto Rico may soon succeed in obtaining their independence. Citizens' initiatives have stopped highways and nuclear reactors and have made possible the legal use of non-Western forms of medicine such as acupuncture. The freedom of society increases as the restrictions imposed on its traditions are removed.

Note how this answer and the attitude that leads up to it differ from the answer and the attitude of rationalists. Rationalists raise questions such as: 'We want to judge the institutions of our societies; we need standards to do that. What are the correct standards? How do we find them?' In asking the question they assume that there is a problem; that any problem *they* perceive is a problem for everyone; and that they are the right people to solve such problems. *They simply take it for granted that their own traditions of standard construction and standard rejection are the only traditions that count.* We have an elitist answer; we do not have a democratic answer.

But the medicine men of central African tribes have no trouble forming an opinion about 'scientific' medicine: they let Western physicians explain their business, they consider the matter, they accept certain forms of treatment and reject others. They have standards and they know how to use them, even in unusual circumstances. Women, trusting more in the regenerative powers of Nature than in the male presumption that sickness is a malfunction that can and must be corrected by scientific tinkering have found their own ways of dealing with a great variety of disorders. And so on. The fact that rationalists have mental blanks certainly does not mean that everybody has.

Alien traditions, it is replied, may have answers where Western intellectuals have problems, but these answers cannot be taken seriously. Science, technology, medicine and other institutions that have been developed in the West are better than alternatives because they have *results*. This is why their problems are real problems which must be taken seriously by everybody. This reply which many philosophers accept without hesitation has no rational basis. The sciences, it is said, are uniformly better than all alternatives – but where is the evidence to support this claim? Where, for example, are the control groups which show the uniform (and not only the occasional) superiority of Western scientific medicine over the medicine of the *Nei Ching*? Or over Hopi medicine? Such control groups need patients that have been treated in the Hopi manner, or in the Chinese manner using Hopi experts and experts in traditional Chinese medicine (rather than Western physicians who have learned the one or the other exotic trick and now already regard themselves as 'experts' in the alien art); but the requested procedures are often against the law and are at any rate frowned upon and sabotaged by the medical socities. Secondly, the reply assumes what is to be shown viz. standards that make the results of science worthwhile. But a mystic who can leave his material body and meet God Himself will hardly be impressed by the fact that thousands of people, using billions of dollars of tax money succeeded in putting two wrapped-up bodies on a hot and dried out stone, the moon, and he will deplore the decrease and almost complete destruction of man's spiritual abilities that is a result of the materialistic-scientific climate of our times. One may of course ridicule such an observation, but one cannot remove it by using the argument from scientific success. Difference in standards and values plays an even greater role in medicine: Western 'scientific' medicine aims at smooth functioning of the body-machine no matter what its feelings or its aesthetic appearance; other forms of medicine are interested in feelings, intuitive abilities, special achievements (prophecy; shahmanism) that cannot be measured in materialistic terms.

Another objection to a democratic relativism is that we live in a scientific age and have to adapt to it. The reply is, first, that this is not true – science is by no means omnipresent – and, second, that even its omnipresence cannot be regarded as an argument for acceptance: if a country is invaded by locusts then it is useful to study their habits but it would be quite unreasonable to turn them into national deities.

The elitism attacked, says a further argument, does no harm, for today everybody can in principle become a member of the elite: everybody can become a scientist, a politician, a Great Thinker, even the President of a University. However one can become a member of the elite only if one adopts its ideology and its habits: equality, equality of women and 'racial' equality included, does not mean equality of traditions; *it means equality of access to one particular tradition*, the tradition of the White Man. White

Liberals supporting the demand for equality have opened the promised land, but it is a promised land built after their own specifications, filled with their own favourite playthings and accessible only in accordance with their requirements (consider the importance of 'intelligence' tests for access to all sorts of activities).[58]

The most effective move against a democratic relativism – for we can hardly speak of an argument – is emotional blackmail or, more correctly, *slander*. For example, many critics raise the spectre of racism, Auschwitz, terrorism, chaos. But democratic relativism denies the right of traditions to impose their form of life on others and therefore recommends the protection of traditions from outside interference. Hopi medicine will be protected from Western medico-fascism just as Jews will be protected from the political fascism of the anti-semites. Nor is the fear of chaos justified: the traditions whose independence we want to protect usually are much stricter towards their members than is the protecting mechanism towards the protected traditions. The belief that the institutions of a free society should protect the individual and not traditions is closely connected with the liberal belief that individuals can exist and have properties worth protecting independently of all traditions. The belief is correct to a certain extent: already a foetus has individuality, it reacts towards its surroundings, it contains the possibilities for a rich and rewarding life. What is not correct is the assumption that preservation of these possibilities is a basic value never to be overruled. Not even liberals make this assumption a basis of their creed (not all liberals are pacifists). Besides, a foetus is not a fully fledged human being; (s)he needs a tradition to become that, and so traditions do become the prime elements of society. Of course there will be cases when the state rightfully interferes even with the internal business of the traditions it contains (e.g. spreading of infectious diseases) for like every rule the rules of democratic relativism have exceptions. The point is that in a democracy the nature and the placement of the exceptions is determined by specially elected groups of citizens and not by experts and that these groups will choose a democratic relativism as the basis on which the exceptions are imposed. The problem of education, however (people may stay in execrable traditions because they do not know better, hence we need a uniform and universal education) provides an argument *against* the status quo, not *for* it. Hardly any adherent of science, scientific medicine, rational procedures has chosen this form of life from among a variety of alternatives; the scientific point of view was imposed by 'education', not chosen,[59] and

[58] Only a few radicals have noticed this restriction. Thus women liberationists fight for the right of women to participate in male manias and only a few women are critical of these manias themselves.

[59] According to Kant, enlightenment occurs when people leave the stage of self-inflicted immaturity. Using this definition we can describe the development since the eighteenth century by saying that immaturity with respect to the churches was replaced by immaturity

the groups who want to leave the fold and return to more traditional forms of life do so in full knowledge of the splendours they are leaving behind: they have tasted the bouquet of scientific rationalism and have found it wanting. We see that neither arguments nor moral pressures can *remove* the democratic relativism that was proposed at the beginning of the present section. And there are lots of arguments *in its favour*.[60]

The first argument, which I have already mentioned, is one of *rights*. *People have the right to live as they see fit.* If there exists a tradition that has religious reasons for rejecting certain forms of medical treatment (some tribes in central Africa do not want to be X-rayed because they do not want their internal organs exposed to view) then no institution should be permitted to force it to accept these forms. Conversely, if there exists a tradition that uses treatment contrary to the ideas of Western medicine, then no institution should be permitted to force it to reject these forms, or to put them at a disadvantage (e.g. no health insurance, or no paid sick leave). Science and rationalism in this view are instruments put at the disposal of the people *to be used by them as they see fit*; they are *not* necessary conditions of rationality, or citizenship, or life. Scientists are salesmen of ideas and gadgets, they are not judges of truth and falsehood. Nor are they high priests of right living. I have already said that there are exceptions to this rule as there are to any rule. The point is that in a democracy these exceptions are dealt with by democratic councils and that the councils take democratic relativism as their starting point.

A second argument in favour of a democratic relativism is closely connected with Mill's arguments for proliferation. A society that contains many traditions side by side has much better means of judging each single tradition than a monistic society. It enhances both the equality of the traditions and the maturity of its citizens. We can learn a lot from 'primitive' tribes about care for the aged, treatment of 'criminal' elements, treatment of (behavioural) aberrations, we can observe the advantages of direct knowledge over an 'objective' account that approaches its object in a standardized and severely alienating way. For Montaigne and his followers in the Enlightenment a study of savage cultures provided not only valuable contributions to the understanding of man, it also put a mirror in front of 'civilized' man and exposed his shortcomings and vices. Today a look at the lives of independent *women* can show us the barbarism that characterizes so much of our *man*-made societies. Democratic relativism makes these contrasts stand out and so enables everybody to learn from them in her (his) own way.

with respect to science and rationalism. Enlightenment is as distant today as it was in the sixteenth century.

[60] A more detailed version of these arguments is found in part 2, ch. 3 of my *Erkenntnis für Freie Menschen*, 2nd edition (Frankfurt, 1980).

A third argument follows directly from the second. Scientific views are not only incomplete – they omit important phenomena – they are often erroneous right in the centre of their competence. Routine arguments and routine procedures are based on assumptions which are inaccessible to the research of the time and often turn out to be either false or nonsensical. Examples are the views on space, time and reality in eighteenth- and nineteenth-century physics and astronomy, the materialism of most medical researchers today, the crude empiricism that guided much of seventeenth- and eighteenth-century science and influenced even the debates about the Darwinian theory. These views are essential parts of important research traditions, but only a few practitioners are aware of them and can talk about them in an intelligent manner. And yet scientists show increased belligerence when such views are attacked and they put the whole weight of their authority behind ideas they can neither formulate nor defend; we need only consider how vigorously some scientists defend a rather naive form of empiricism without being able to say what facts are and why anyone should take them seriously. The lesson to be learned from this phenomenon is that *fundamental debates between traditions are debates between laymen which can and should be settled by no higher authority than again the authority of laymen, i.e. democratic councils.*

The situation just described becomes important in cases where the belief in the soundness of a view or an enterprise has led to institutional (and not only to intellectual) measures against alternatives. 'Scientific' medicine is a good example. It is not a monolithic entity; it contains many departments, schools, ideas, procedures and a good deal of dissension. Nevertheless there are some widespread assumptions which influence research at many points but which are never subjected to criticism. One is the assumption that illnesses are due to material processes which can be localized and identified as to their chemico-physiological nature, and that the proper treatment consists in removing them either by drugs, or by surgery (complex methods of surgery such as laser surgery included). Now we may ask whether the problems of scientific medicine – and there are many – have something to do with this assumption or whether their origin lies elsewhere. It is well known that deathrates in hospitals go down when doctors are on strike; is this due to the incompetence of the physicians, or does it show a basic fault in the theoretical structure that guides their actions? We know that cancer research absorbs lots of money and has few results.[61] Is this due to the fact that cancer researchers are mainly interested in theory, not in healing, or does it indicate a basic failing of the theories used? We do not know. To find out we must make the basic assumptions (e.g. materialism) visible and examine them in a more direct manner. To examine them in a more direct

[61] Cf. Daniel Greenberg, 'The "War on Cancer": Official Fiction and Health Facts' *Science and Government Reports* IV (1 Dec. 1974).

manner we have to compare the results of scientific medicine with the results of forms of medicine based on entirely different principles. Democratic relativism permits and protects the practice of such different forms of medicine.[62] It makes the needed comparisons possible. *Democratic relativism*, therefore, not onls *supports a right, it is also a most useful research instrument for any tradition that accepts it.*

Democratic relativism, then, is a fine thing, but how are we going to introduce it? And how are we going to keep the various traditions in place and prevent them from overwhelming each other by force as Western conquerors once overwhelmed old cultures? The answer is that the necessary institutions *already exist*: almost all traditions are part of societies with a firmly entrenched protective machinery. The question is therefore not how to *construct* such a machinery, the question is how *to loosen it up* and to detatch it from the traditions that are now using it exclusively for their purposes; for example, how to separate state and science. The answer to *this* question is that the methods employed cannot be discussed independently of the tradition that wants to achieve equality and the situation it finds itself in. The democratic relativism I have discussed will not be imposed '*from above*', by a gang of radical intellectuals, it will be realized *from within*, by those who *want* to become independent, and in the manner *they* find most suitable (if they are a lazy bunch, then they will move *very* slowly, and with long periods of rest in between their political interventions). What counts are not intellectual schemes, but the wishes of those who want change. Or, to use a catchy slogan: *Citizens' initiatives instead of philosophy!* Details are found in my *SFS* and in the much improved German version: *Erkenntnis für Freie Menschen*, 2nd, revised edition (Frankfurt, 1980).

[62] The objection that people must be protected cannot be raised at this stage; after all, we do not yet know, *before* the comparison, what they must be protected from. It may well turn out that we must protect them from mutilation by 'scientific' body plumbers.

2
Classical empiricism

1. In the present paper I want to describe certain features of post-Galilean, or 'classical' science that deserve greater attention than has been given them so far. These features are summarized in the following three points.

(i) The practice of post-Galilean science is *critical* in the sense that it allows for the revision of any part of it, however fundamental and however close to 'experience'. Resistance must of course occasionally be overcome, but the resistance is never strong enough to stabilize completely some particular piece of knowledge.

(ii) This critical practice is accompanied by a *dogmatic ideology*. The ideology admits that science may contain hypothetical parts. But it is emphasized that such parts are preliminary, that they will either disappear as the result of further research, or will turn into trustworthy theories. Moreover, it is assumed that all theories rest on one and the same stable foundation, *experience*. It is experience which supports, and gives content to, our ideas without itself being in need of support and interpretation.

(iii) Thus we have on the one hand the assumption of a stable foundation while we are on the other hand engaged in an activity that prevents such a foundation from ever coming into existence. Now the peculiarity of post-Galilean 'classical' science I want to describe consists in the manner in which this abyss between ideology and practice is bridged: first, experience is identified as that part of a newly conceived hypothesis that can most readily be illustrated by simple and eye-catching procedures. Secondly, the experience so defined is made solid by the success (which may have been brought about with the help of *ad hoc* assumptions) of the hypothesis it illustrates as well as by the vividness of the illustrating examples. Thirdly, it is given the appearance of *stability* through a method of interpretation that aims at, and succeeds in, concealing all change. The aim is reached by concentrating on the illustrations themselves rather than on the role they play in a particular theory (the same picture, after all, can illustrate very different things). I use the term *classical empiricism* to describe this fascinating, tortuous, schizophrenic combination of a conservative ideology and a progressive practice. The most outstanding practitioner of classical empiricism is Newton.

2. Classical empiricism differs both from the empiricism of Aristotle and from the empiricism of Bacon.

The demand that we base knowledge upon experience makes excellent sense in the *Aristotelian philosophy* where experience is defined as the sum total of what is observed under normal circumstances (bright daylight; senses in good order; undisturbed and alert observer) and what is then described in some ordinary idiom that is understood by all.[1] Aristotelian empiricism, as a matter of fact, is the only empiricism that is both clear – one knows what kind of thing experience is supposed to be – and *rational* – one can give reasons why experience is stable and why it serves so well as a foundation of knowledge.

For example, one can say that experience is stable because human nature (under normal conditions) is stable. Even a slave perceives the world as his master does. Or one can say that experience is trustworthy because normal man (man without instruments to becloud his senses and special doctrines to becloud his mind) and the universe are adapted to each other; they are in harmony.

This rational context which enables us to understand the Aristotelian doctrine and which also provides a starting point of discussion is eliminated by the 'enlightenment' of the sixteenth and seventeenth centuries.

3. It is characteristic of this enlightenment that it constantly mentions new and undiluted foundations of knowledge and of the faith while at the same time making it impossible ever to *identify* these foundations and to build on them. (This corresponds to (ii) and (i) of section 1.)

Thus Luther and Calvin (1) declare Holy Scripture to be the foundation of all religion. This is the new Protestant rule of faith from which everything else is supposed to proceed. But we are also urged (2) to put aside, and never to use what cannot be justified by this rule. Now this second step clearly voids the first or, to express it differently, the Protestant rule of faith as expressed in (1) and restricted in (2) is *logically vacuous*. The argument, briefly, is as follows.[2]

[1] G. E. L. Owen, 'ΤΙΘΕΝΑΙ ΤΑ ΦΑΙΝΟΜΕΝΑ,' *Aristote et les problèmes de la méthode* (Louvain, 1961), 83–103. Reprinted in *Aristotle*, ed. Moravcsik (New York, 1967), 167–90. See also Aristotle, *De Anima, De Sensu, Anal. Post.*, and parts of *De Part. Anim.*

[2] (*a*) For the Protestant rule of faith see Martin Luther, *The Babylonian Captivity of the Church* quoted from Henry Bettenson (ed.) *Documents of the Christian Church* (Oxford, 1947), 280: 'For what is asserted without the authority of Scripture and of proven revelation may be held as an opinion, but there is no obligation to believe in it.' For an interesting addition see the following report of Luther's encounter with Nicholas Storch (falsely called 'Mark' in the report) as quoted in Preserved Smith, *Life and Letters of Martin Luther* (Boston, 1911), 150: 'In 1522 Mark Storch came to me with sweet seductive words to lay his doctrines before me. As he presumed to teach things not in Scripture I said to him: "I will not agree with that part of your doctrine unsupported by Scripture *unless you work miracles to prove it*" . . . He said: "You shall see miracles in seven years" . . .' (my italics).

Despite his insistence upon foundations and upon the purity of faith, and despite his

(*a*) The rule does not provide any means of *identifying* scripture (no version of scripture contains a passage to the effect that 'the preceding . . . and the following . . . pages are Scripture'). We are told what the basis of

frequent violence in writing, Luther seems to have objected consistently to the use of force; he was also quite tolerant towards deviations for which a good reason could be given. 'I have gathered from the writings of these people' he writes to Frederick of Saxony (letter of July 1524, quoted from Smith, *Life and Letters*, 151) discussing the action of some contemporary revolutionaries, 'that this same spirit will not be satisfied to make converts by word only, but intends to betake himself to arms and set himself with power against the government, and forthwith to raise riot. Here Satan lets the cat out of the bag. What will this spirit do when he has won the support of the mob? Truly here at Wittenberg I have heared from the same spirit that this business must be carried through with the sword . . . It is a bad spirit which shows no other fruit than burning churches, cloisters, and images, for the worst rascals on earth can do as much . . . It is (also) a bad spirit which dares not give an answer . . . for I, poor, miserable man, did not so act in my doctrine . . . I went to Leipzig to debate before a hostile audience . . . If they will do more than propagate their doctrines by word, if they attempt force, your Graces should say: we gladly allow any one to teach by the word, that the right doctrine may be preserved; but draw not the sword, which is ours . . . they are not Christians who would go beyond the word and appeal to force, even if they boast that they are full of holy spirits.' In his work *Against the Heavenly Prophets of Images and the Sacrament*, the first part of which appeared late in December 1524, he defends pictures 'as a help to the faith of the ignorant' (Smith, 156). 'These prophets' he continues, criticizing such radicals as Carlstedt, Muenzer, Storch 'teach that the reform of Christendom should start with a slaughter of the godless, that they themselves may be lords of the earth . . . [But] those who preach murder can have no other origin than the devil himself.' 'On August 22, 1524' writes Smith (154) describing a meeting between Luther and Carlstedt 'the two had a conference in Jena and parted with a friendly agreement to differ. "The more ably you attack me" said Luther "the better I shall like it", and gave his old colleague a gold gulden as a sign that he was free to advance what opinions he liked so long as they were supported by argument only and not by violence.' Even after Carlstedt had preached sedition, and spread death and destruction, Luther would still give him shelter and write to the Elector in his defence. There were many admirable qualities in Luther and he was strong enough to withstand the consequences of his own mischievous rule of faith. But irrational rules share with the institution of tyranny this one disadvantage: they may work wonders in the hands of men strong enough to go their own ways; but they create havoc in the hands of almost everyone else. This becomes clear at once when turning to Calvin.

Calvin showed much less tolerance (example: the execution of Servetus). He was much clearer, much more consistent, and therefore much more terrible in his unflinching pursuit of what he thought to be the proper faith. (This, by the way, is one of the reasons why I would prefer lukewarm defenders of inconsistent doctrines to serious opponents of clear, precise and, above all, consistent views.) The reader interested in Calvin's formulation of the rule of faith should turn to ch. 7 of the *Institutes of the Christian Faith*, in J. T. McNeill (ed.) *On the Christian Faith* (Indianapolis, 1958), 19ff, as well as to R. H. Popkin, *The History of Skepticism from Erasmus to Descartes* (New York, 1964). Professor Popkin's book has been a great help to me, both in a course which I gave in summer 1966 at the University of California in Berkeley (and which led me away from the philosophy of science into church history) and in the preparation of the present paper.

Both Luther and Calvin are anticipated by (and consciously follow) St Paul who writes (Colossians, ii): 'Beware lest any man spoil you through philosophy and vain deceit derived from the *traditions* of men, conforming to the rudiments of the world, and not to Christ.'

(*b*) The argument in the text is due to the Jesuit Father François Véron of La Flèche where Descartes received his early education. It is summarized in Popkin, *History of Skepticism*, 72ff where the reader will also find an excellent description of Véron's method and of his impact. The argument anticipates, and excels in clarity and conciseness, all the subsequent criticisms of fundamentalist doctrines up to and including Wittgenstein.

the right faith ought to be; but we do not receive any indication as to how we can find this basis among the many books and tales in existence.

(*b*) Given scripture we do not know how to *interpret* it (no version of scripture contains a grammar and a dictionary of the language in which it is written. Such a grammar and such a dictionary are of course available, and often unnecessary; for example, they are unnecessary when we understand the language of the bible. But then our traditional understanding of a particular language is *added* to scripture whereas the rule of faith, and especially the second principle enounced above wants scripture to be the only authority. We see how much more reasonable *and human* the Roman position has been).

(*c*) Given scripture and a certain reading of it we have no means of *deriving consequences* (no version of scripture contains a logic or a more general system for the production of statements on the basis of other statements). Even if we recognize the basis of our faith, and even if we know how to interpret it, still we have no means of going beyond it, not in the simplest matter. For example, we cannot apply it to contemporary problems.

4. Now it is interesting to see that these objections which were first put forth by a Jesuit Father (see n.2(*b*)) and which have never been superseded in clarity and conciseness, apply point for point to the Baconian philosophy, the second great fundamentalist doctrine of the seventeenth century.

The 'rule of faith' of Baconian science is experience. Again this rule does not allow us (*a*) to *identify* experience (see the corresponding letters in the preceding section). Just as it was taken for granted by the defenders of the new faith that scripture was known to everyone, in the very same manner the identity of experience is now assumed to be known without the shadow of a doubt. This was permissible as long as one defined experience in the Aristotelian manner, comprising all those things and processes which are noticed under normal circumstances, with one's senses in good order, and which are then *described* in some ordinary idiom; that is, as long as one could make use of some *tradition* (cf. section 2). However, just such a use of tradition (or of preconceived opinion, as tradition was soon called) is to be avoided. This is the essence of the Baconian creed. But with it the only means of identifying experience in a rational manner disappears also. The first objection of the previous section applies.

Nor are we able (*b*) to determine what experience tells us. Experience, taken by itself, is mute. It does not provide any means of establishing a connection with a language unless one already includes in it some elementary linguistic rules, i.e. unless one again refers to tradition. Wittgenstein saw this point and has expressed it more or less clearly. He saw that whenever we observe and report observations we make use of traditional elements ('forms of life'). However, he has failed to indicate how traditions can be improved and has even created the impression that such improvement is impossible.[3]

Finally, (*c*), there is no way of obtaining the complex theories which were soon used to support the empiricist creed (and which in turn received support from it) even if one makes the additional assumptions that experimental statements are singular statements and that means of deduction are readily available. This point was made most forcefully by Berkeley (*Principles of Human Knowledge*, section 107), and by Hume.

It is rather interesting to examine the similarities between the theories of Protestantism and of Baconian empiricism. These similarities are expressed not only in the *structure* of the respective doctrines, but even in the phrases which are used to direct attention to the respective bases (scripture; experience): reverence is demanded of both of them, success and a clear view of an all-embracing entity (God; nature) is promised in both cases, and in almost the same exalted terms.[4] A detailed description of such phenomena is a challenging task for the historian of ideas.

Now what interests us in the present paper is not only the (Protestant; Baconian) ideology, but also the way in which this ideology is applied in practice. It is the practical application of the (Protestant and empiricist) rule of faith that will return us to (iii) of section 1 and serve as an introduction into Newton's philosophy. In a nutshell this practical application can be described by saying that the rule of faith, *although logically vacuous, is by no means psychologically vacuous.* We first illustrate this feature in the case of Protestantism, especially of the Calvinist variety.

5. To do this we must remind ourselves that the rule is not introduced into a vacuum. It is introduced, and taught, in a community (such as the community of Calvinists in Geneva in the sixteenth century) which *is already committed to* a certain doctrine. Children are educated by Calvinist parents, in Calvinist schools where religious instruction plays a fundamental role and pervades all subjects; they are encouraged when they say the 'right things', punished when they 'do wrong'. From their most tender years they are part of a life all aspects of which are guided by religious considerations. The language they learn is permeated with religious sentiment and is structured accordingly. Bible passages ring in their ears, for they are read and pronounced on every and any occasion. Their knowledge of the bible is constantly tested, by examinations at school, in conversations with others; and mistakes, i.e. deviations from the accepted point of view, from the

[3] For this point cf. section 15 of my 'Problems of Empiricism', ch. 3 in *AM*.

[4] 'We have now treated of each kind of idols, and their qualities, all of which must be abjured and renounced with firm and solemn resolution, and the understanding must be completely freed and cleared of them, so that the access to the kingdom of man, which is founded on the sciences, may resemble that to the kingdom of heaven, where no admission is conceded except to children' (Francis Bacon, *Novum Organum*, Aphorism 68).

'party line' are corrected at once. Having been prepared in this fashion they will of course interpret 'scripture', i.e. the books used in the community, as everyone else does and they will not at all be aware of the logical gap that separates this common faith from the 'word of God' as defined by the rule. The believers will simply perceive in the lines of scripture the very same god they have been taught to revere. Now the rule of faith, which in Calvin's version refers to an 'intuitive perception of God himself' (see n.2(*a*)), is wide and indefinite enough to allow this act of recognition to become a foundation of the faith. On the other hand, the indoctrination is specific enough to identify this act as the *sole* foundation of the faith. Thus the *psychological* result of indoctrination becomes the *theological* basis of the doctrine taught[5]

[5] This is a very familiar procedure. It occurs, in various forms, at almost every stage of the history of the Church (of science, of law, and what have you). Thus already the *selection of the canon* was determined by the doctrines held at the time. Alfred Wikenhaeuser, in his *Einleitung in das Neue Testament* (Herder, 1961), 57, describes the situation by saying that 'the text of the New Testament has been preserved not critically, but dogmatically' which means, when put into plain English, that the historically available material was changed in order to fit an already existing faith: *this faith creates its own sources*. The influence of the popular faith may also be seen from the fact that St Jerome, when translating the Vulgata, kept as closely as possible to the widely read Itala as 'numerous and far going changes in the generally known wording would have only created resentment' (*Letter to Damasus* as paraphrased in Wikenhaeuser, 75). 'The criticism which Luther applied to the received canon is (still) bolder and much radical. It is not a historical criticism, but is dogmatically, or religiously oriented. Luther made his interpretation of Paulus the measure by which everything allegedly divine or holy would have to be tested. The 'true and certain main books' of the New Testament are according to him Romans and Galateans, and of the gospels the gospel according to John: next come the other gospels, the synoptics, because they contain so few genuine words of Jesus; in addition he mentions a third group which he reprimands severely "as it does not advance Christianity"' (Wikenhaeuser, 44).

The *early history of the canon* is described in great detail by A. Harnack, *History of Dogma*, trans. N. Buchanan (New York, 1961) II, 43ff. I quote a few excerpts: 'The canon emerges quite suddenly in an allusion of Melito of Sardis preserved by Eusebius . . . in the works of Irenaeus and Tertullian; and in the socalled Muratorian fragment. There is no direct account of its origin, and scarcely any indirect; yet it already appears as something to all intents and purposes finished and complete. Moreover it emerges in the same ecclesiastical district where we first were able to show the existence of the apostolic *regula fidei*.' The principle of selection used by the Church 'was to reject as spurious all writings, bearing the names of the Apostles, that contained anything contrary to Christian commonsense, that is, to the rule of faith – hence admission was refused to all books in which the God of the old Testament, his creation, etc. appeared to be deprecated – and to exclude all recensions of apostolic writings that seemed to endanger the Old Testament and the Monarchy of God.' Again already existing views are made the measure of the alleged basis.

The origin of the rule of faith itself is even more interesting.

From the very beginning, but especially after the rise of Gnosticism, the Church tried to bring about a synthesis of the elements of Christianity (scripture; sacraments; organization; baptismal symbol) that would *separate* it from the flood of contemporary speculation and would *preserve* and *emphasize* its unique character. It was necessary, for this purpose, both to *deny* the main tenets of Gnosticism (and of later heresies); and to show that this denial was not *ad hoc*, but was implied by the very essence of the Christian faith. With this aim in mind, Gnosticism was declared to be obviously inconsistent with the content of the simple and quite untheoretical prayers in which the believers expressed their faith (for an example of such a prayer see Bettenson, *Documents*, 90ff). There is no doubt that the spiritual content of the prayers was thereby drastically changed. Very soon the form began to change also; terms of

and further strengthens belief in it (cf. (iii) of section 1). The circle is closed; the 'chosen children of God', as a follower of Calvin (Beza) calls the faithful, are separated from the 'castaways', are made aware of their mission, and are encouraged in their faith.

It is clear that this marvellous method which turns an obvious weakness (the irrational character of the rule of faith) into overpowering strength can work only if the rule does not introduce new and unknown elements. The logical vacuity of the rule is precisely what is needed if it is to fulfil its most important practical function: *to reinforce an already existing faith.* The results of historical research, to mention but one point on which the Romans differed

speculative philosophy intruded so that the prayers which originally had expressed the faith of the faithful in a very immediate way finally became summaries of a steadily growing *body of doctrine*. As an example one should consider the creed of Nicaea (Bettenson, *Documents*, 36) and compare it with the simpler early formulations: 'We believe in *one* God the Father All-Sovereign, Maker of all things *visible* [against the demiurge of the Gnostics] and invisible; and in one Lord Jesus Christ, the Son of God, *begotten of the Father* [against adoptianism], *only* begotten [against the Gnostic hierarchy], that is, of the *substance* of the Father [against Arianism], God of God, Light of Light, true God of true God, *begotten, not made*. . . .' The whole process is described by Harnack, II, 26ff in the following words 'What was needed [in order to refute the Gnostics who recognized the creed of the old Church "since they already possessed the art of explaining a given text in whatever way they choose'] was an apostolic creed *definitely interpreted* [my italics]; for it was only by the aid of a definite interpretation that the creed could be used to repel the Gnostic speculations and the Marcionite conception of Christianity.

'In this state of matters the Church of Rome, the proceedings of which are known to us through Irenaeus and Tertullian, took, with regard to the fixed Roman baptismal confession ascribed to the apostles, the following step: *the antignostic interpretation* required by the necessities of the times *was proclaimed as its self-evident content*; the confession, thus explained, was designated as the "Catholic faith" (*fides catholica*), i.e. the rule of truth for the faith; and its acceptance was made the test of adherence to the Roman Church as well as to the general confederation of Christendom . . . What the Roman community accomplished practically was theoretically established by Irenaeus and Tertullian. The former proclaimed the baptismal confession, definitely interpreted and expressed in an antignostic form, to be the apostolic rule of truth (*regula veritatis*), and tried to prove it so. He based his demonstration on the theory that this series of doctrines embodied the faith of the churches founded by the Apostles, and that these communities had always preserved the apostolic teaching unchanged.' For Irenaeus see *Adversus Haereses*, III/iii.1 as well as IV/xxvi.2. For Tertullian *De Praescriptione Haereticorum*, xx.

I mention these parallels in church history mainly as an aid for the better understanding of those features of science in which we are here interested. In religion such features are clearly visible for they are regarded as being of paramount importance. A study of religion will therefore school our eyes and make them prepared for the darkness we are about to encounter. The reasons for the better visibility are to be sought in the wider interests of the churchman and of the church-historian. Scientists and scientific historians are *specialists* and their interests are rather narrowly defined. They will therefore be inclined either to overlook, or consciously abstract from, phenomena which exhibit certain general tendencies of man. Theologians are specialists, too, but as their subject matter is the whole of man they will be much more sensitive towards the changes of doctrine with which we are here concerned. This sensitivity is clearly exhibited by St Jerome (see above) who in this way makes us understand similar phenomena in the sciences.

For a more detailed examination of the effect of Protestantism on science, see S. F. Mason, 'The Scientific Revolution and the Protestant Reformation' *Annls Sci.*, 9 (1953), 64–87, 154–75.

from the Protestants, cannot be foreseen. They may contradict the doctrine and endanger the loyalty of its adherents. It is therefore the wish to preserve party lines that makes it so essential to use a vacuous rule and, to mention another feature of Protestantism which will not be discussed in the present paper, to keep different subjects apart.

Now this process, this use of a vacuous rule of faith in the practical defence of ideas that one wants to preserve, is also one of the most characteristic features of classical empiricism. That devils, gods, witches, have all been defended on empirical grounds is well known. However, for the present purpose a brief discussion of Newton's theory of colours is more appropriate.[6]

6. When speaking of Newton's theory of colours I am referring to what one might call the *ray theory of light* according to which light consists of rays of different refrangibility and different colour whose inherent properties are not changed, either by refraction, or by reflection, or by any other process, and which produce colours either singly, or by mixture. This *theory* is to be distinguished from the *corpuscle hypothesis* which Newton also held, and from the further hypothesis that light is a substance, which he occasionally presented as a direct and unique consequence of phenomena.[7]

[6] In what follows I shall use the *text* to give an abbreviated and somewhat dramatic account of Newton's philosophy while I shall use the footnotes for bringing in details, sources, and for establishing connections.

Up till now the historical literature concerning Newton has been full of the kind of dogmatism he himself tried to put over on his contemporaries and successors. (There were a few exceptions such as Goethe, but they were mostly regarded as cranks. However, German scientists were put in a difficult position by their joint veneration of their chief poetic father figure and of what they thought was proper scientific method.) For example, it has been taken for granted, almost universally, that Newton established the nature of white light. This situation is now at last changed by the appearance of Dr A. I. Sabra's *Theories of Light from Descartes to Newton* (London, 1967). This, as far as I can see, is the first consistently critical analysis of Newton's optics and the first account that explicitly considers his methodology and its role in the theory of colours. The dominance of Newton over the *historians* has at last come to an end (in *physics* this dominance was terminated over a century ago, though there are still some textbooks for which these developments are practically non-existent).

[7] For the ray theory see Newton's first paper on light and colours, *Phil. Trans. R. Soc. Lond.* A no. 80 (1671), 3085. This paper and other papers of Newton are quoted from I. B. Cohen (ed.), *Isaac Newton's Papers and Letters on Natural Philosophy* (Cambridge, 1958), 57ff. For the idea that light is a substance see R. S. Westfall, 'The Development of Newton's Theory of Colours', *Isis*, 53 (1962), 352. The remainder of Westfall's paper is somewhat too uncritical towards Newton's achievements.

Newton's *methodological beliefs* and prescriptions may be summarized as follows. It is assumed that general propositions, also called *theories* can be derived from *phenomena* (see below) 'concluding positively and directly', and not only 'from a confutation of contrary suppositions' (*Phil. Trans. R. Soc. Lond.* A, 8 July 1672), 4004; Cohen, 93); and that they can be 'proved' in this manner. This implies a distinction between phenomena (or facts) and theories on the one hand, and hypotheses on the other. Theories are related to facts and phenomena in the unique way just described. Hypotheses are invented (if at all) only *afterwards*, for '*explaining* the properties of things'; they are not 'assumed in determining them'

The problems of the ray theory as well as the later history of that theory (and of the logically quite different theory of gravitation) have much in common with contemporary argument in microphysics. So great is the similarity that for each relevant feature that needs discussion in the one case a corresponding feature can be found in the other. One might almost

(Reply to Pardies' second letter, Cohen, 106). They must never contradict phenomena or theories. The nature of things, insofar as it shows itself in experiment, and their behaviour is settled first, and in a unique way. Reasons for this behaviour and this nature are given later, and with the help of hypotheses. (Note how little this last feature differs from the much more 'modern' expositions of, say, Nagel, where high level theories must agree with low level theories which have been 'established in some area of inquiry'. See Ernest Nagel, *The Structure of Science* (New York, 1961), 338; see also my exposition and criticism of this book in ch. 3. The only difference is one of length. Newton expressed himself briefly, and to the point.) No hypotheses are needed if the problem is to establish a theory, or a phenomenon; 'to dispute about *hypotheses*... is [then] beside the business at hand' (Cohen, 123). It is also implied that while hypotheses are to be judged by theories and phenomena 'and those to be rejected which cannot be reconciled with the phenomena' (108; cf. Newton's second paper where he asserts that he 'gave a reason why all allowable hypotheses [on light] should be conformable to [his] theories, 178) theories can be criticized only by 'showing the insufficiency of [the] experiment used... or by assigning the flaws and defects in [the] conclusions drawn from them' (94). (In effect this is again identical with the method entailed by the present theories of reduction and explanation now presented, e.g. by Nagel in *The Structure of Science*.) Newton is very emphatic about this *asymmetry* between experimental results and theories on the one hand, hypotheses and associated speculation on the other; he finds it 'necessary' to start the labour of science by 'lay[ing] aside all hypotheses' (Cohen, 106) and he emphasizes that even later on, considerations of truth and reality must be related to experiment and theories deduced from them and must not be made dependent on the consideration of alternative hypotheses: 'For if the possibility of hypotheses is to be the test of truth and reality of things I see not how certainly can be obtained in any science; since numerous hypotheses may be devised which shall seem to overcome new difficulties' (reply to Pardies' second letter, 106; note that Newton takes the devising of numerous hypotheses 'to be no difficult matter', reply to Huygens, 144).

All these ideas find concise expression in rule 4 of the *Principia*. Newton has rewritten this rule at various times. Some formulations make clear the connection with his earlier point of view: 'In experimental philosophy one is not to argue from hypotheses against propositions drawn by induction from phenomena. For if arguments from hypotheses are admitted against inductions, then the argument of inductions on which all experimental philosophy is founded could always be overthrown by contrary hypotheses. If a certain proposition drawn by induction is not yet sufficiently precise, it must be corrected not by hypotheses, but by the phenomena of nature more fully and more accurately observed.' (See Alexandre Koyré, *Newtonian Studies* (London, 1966), 269; for the final formulation see the text to n.18). The final version of this rule has influenced the development of modern science down to our own day.

Let us now examine the effect of the combined use of this rule, of a general proposition, and of the assumption that the proposition has been inferred from phenomena, 'concluding positively and directly'.

The rule and the philosophy on which it rests allow for two modifications of 'propositions inferred by general induction from phenomena', or of *theories* as Newton calls such propositions. They may be made more precise; and they may be made liable to exceptions. In the first case one adds to their informative content; in the second case one subtracts from it by restricting the domain of application. It is assumed that this domain will not shrink to zero. Such a shrinkage would indicate that there are no phenomena from which to derive the theory – a supposition that is at once refuted by a repetition of the relevant experiments. The concepts of the theory which express these phenomena will therefore remain *in use forever* and so will all those principles which have been 'established in some area of inquiry'. For further details see Sabra, *Theories of Light from Descartes to Newton*, ch. 11.

say that Newton has anticipated all the arguments which are used today for the defence of the Copenhagen Interpretation. Three things are at issue: the value of a theory; the value of the methodology (of the 'rule of faith') on which the theory is allegedly based; the relevance of the methodology and the question of whether it leads to a unique determination of the theory. The interest of the debates at the Royal Society and of the later discussions lies in the way in which these three separate items are made to support each other, so that a solid and almost irrefutable bastion of doctrine arises. One soon agrees that Newton's theory is more detailed than the available alternatives and gives a better account of almost everything that is known about light. The possibility of alternatives is still emphasized, but no one is able to go beyond general suggestions. This considerably reduces the psychological weight of the alternatives. Newton's ability to illustrate the basic principles of his theory by simple and ingenious experiments works in the same direction. But Newton goes one step further and regards the illustrations as an experimental basis.

7. Two identifications are involved in this step. First, the experimental result is identified with what Newton calls a *phenomenon* which is an idealized and generalized description of it that *uses the terms of the theory under review.*[8] The idealizations are again of two kinds. First, the peculiarities of

[8] In the optical papers and also in the *Opticks* the reference to colours is neither to sensations, nor to properties of everyday objects. It is true that difference of sensation is sometimes taken to indicate difference of physical colours: 'Certainly it is much better to believe our senses informing us, that red and yellow are diverse colours' (answer to Hooke, *Phil. Trans. R. Soc. Lond.* A, no. 80 (1671), 5088; Cohen, *Newton's Papers and Letters*, 126). But this is not always the case. A distinction is drawn between subjective and objective colours: 'I speak here of colours insofar as they arise from light; for they appear sometimes from other causes, as when by the power of phantasy we see colours in a dream' (*Opticks* (New York, 1952), 160f; cf. also the explicit specification in proposition 7, theorem 5 of book 1, part 2). Objective colours remain unchanged whereas subjective colours change as a result of mixture: '. . . when the several sorts of rays are mixed, and in crossing pass through the same space, they do not act on one another so as to change each others' colorific qualities . . . but in mixing their actions in the sensorium beget a sensation differing from what either would do apart' (*Opticks*, 159). It is also emphasized in Newton's answer to Hooke (Cohen, 127) and in his reply to Huyghens (Cohen, 139) that visual colours may have 'a double origin, the same colours to sense being in some case compounded, and in others uncompounded'. The possibility of 'a new kind of white' is considered which, though giving the same impression to the eye as normal sunlight, would yet 'have different properties from it' when examined with the prism. Furthermore 'there are as many simple . . . colours as degrees of refrangibility' (Cohen, 140) that is, Aleph One, whereas the number of visible colours is finite, and a small number at that (the *SCC-NBS Method of Designing Colours* National Bureau of Standards Circular 553 (Washington, 1955) uses 266 categories with not more than 100 and sometimes as few as one item in each category. Meanings and translations are given for 7,500 colour names.) We can even give a quite interesting *physical* reason for this disparity between visual colours and physical colours. It lies in the fact that it is impossible to separate perfectly the 'primary' or 'homogeneal' colours by experiment. This proof, which makes use of the wave theory of light and of the fact that the simple harmonic waves which represent the primary colours in this theory are of infinite duration, consists in pointing out that any spectrum, however detailed,

each single experiment and those features of it which do not allow for an immediate description in terms of the theory are omitted. This is how the 'pyramids' reported by Linus and confirmed by von Helmholtz disappear from sight.[9] (They leave their traces, though, as is seen from the relevant

appears only when the light source is turned on and disappears with it. See R. W. Ditchburn, *Light* (New York, 1963), I, 102. It is shown then that Newton, when referring to colours, does not refer to *sensations*.

Nor are the colours *properties of everyday objects*: '. . . colours are the quality of light, having its rays for their entire and immediate subject . . .' 'Besides, whoever thought any quality to be a heterogenous aggregate, such as light is supposed to be . . .' (Cohen, 57). 'Colours of objects are nothing but a disposition to reflect this or that sort of ray more copiously than the rest; in the rays they are nothing but their dispositions to propagate this or that motion into the sensorium, and in the sensorium they are sensations of those motions under the form of colours' (*Opticks*, 125).

The 'simple' or 'homogeneal' colours which according to Newton are the ultimate constituents of all fields of radiation are therefore strictly speaking *unobservable*: it is not possible to produce them in a physically pure form (cf. Newton's own *caveat* in Cohen, 59, as well as Sir David Brewster's discussion in his *Memoir of the Life, Writings, and Discoveries of Sir Isaac Newton*, (Edinburgh, 1855), I, 116); cf. also the general proof at the end of the last paragraph but one. Nor would it be possible to distinguish them experimentally from an infinite variety of weak mixtures, or disturbances centring around a prominent peak. To present the results of concrete experiments with all their imperfections in a unique form and as *supporting* (and not merely *illustrating*) the ray theory therefore involves idealizing assumptions which bend the concrete and very complex experimental result in the direction of Newton's theory. 'The effect of this "demonstration" writes Sabra (*Theories of Light*, 249) 'is, it must be admitted, almost hypnotic. Nevertheless it is certainly inconclusive'. See also Westfall, *Isis*, 53 (1962), 351f as well as T. S. Kuhn in Cohen, 34f: 'Newton combined a precise and detailed description of his experimental apparatus with an imaginative idealization of his experimental results.' (Cf. also the next two footnotes.) The soporific influence which this method still exerts upon contemporary historians can be seen from some of the reviews of Dr Sabra's book (such as Westfall's).

[9] Linus, in his criticism of Newton, pointed out that the spectrum was never capped by a 'semicircular' end but rather terminated in a 'sharp cone or pyramis' (Cohen, 151). Kuhn points out, quite correctly, that Newton never replied to this criticism. Newton could well afford to do so as he was never quite definite about the semicircularity. In his first paper he writes as follows: '. . . the decay of light was so gradual, that it was difficult to determine justly what was their figure; yet they seemed semi circular' (Cohen, 48). Even in the later *Opticks* he still writes that 'at its sides [the spectrum] was bounded pretty distinctly, but on its ends very confusedly and indistinctly, the light there decaying and vanishing by degrees' (29). His description of what is seen, including the drawings, is therefore unexceptionable. But if this description were regarded as the description of *phenomenon*, that is, as a description of what the *physical light* does under certain circumstances, then Newton's theory would have to be regarded as refuted. The transition from 'what is seen' to the 'phenomenon' in Newton's sense (see n.13) is therefore completely unaccounted for. We know today the reason why pyramids are *seen* even if the *physical light* should happen to terminate in a semicircle (see H. L. F. Helmholtz, *Physiological Optics*, ed. and transl. J. P. C. Southall (Rochester, 1924) I, 173). We may suspect that Newton, who also dealt with the physiology of sight, had an explanation of this kind in mind. However, he did not give the explanation but simply redescribed what he saw in order to turn it into a physically useful phenomenon. And in this redescription he introduced the machinery of the very same theory he wanted to prove. Goethe's question 'for how should it be possible to hope for progress if what is inferred, guessed, or merely believed to be the case can be put over on us as a fact?' addresses itself to this feature of the theory. (See *Farbenlehre*, ed. Ipsen, (Leipzig, 1927), 393.) Dr Sabra (and, of course, Professor Ronchi) apart, I do not know of a single historian of science whose critical sense matches Goethe's in this connection. For Ronchi's view see the quotation at the end of

drawings in the *Opticks*, 29.) This is also the reason why Brewster thought it necessary to assert that the 'homogeneal colours' always contain *all* colours, though in vastly different amounts.[10] Newton was well aware of these matters. He gave detailed instructions for the approximate realization ('exhaustion' as Dingler would have said) of what he considered to be the elementary process of colour separation and he pointed out that success could never be complete.[11] At the same time he described the approximations in a manner suggesting that complete separation *had* been achieved, thus greatly diminishing the distance between 'nature' and his theory.

Secondly, not all experiments are given equal weight. Those whose distance from the theory is minimal and which seem to be a visible expression of its basic principles are preferred to others which do not allow us to read the theory in them at once glance. Thus the different refrangibility of rays of light is demonstrated in the most convincing fashion by the

n.13. Sabra's critical attitude may be explained as a result of his interest in 'pure philosophy' (among other things he attended Sir Karl Popper's seminar at the London School of Economics in 1952, where I first had the pleasure of meeting him).

[10] Brewster, *Memoir of Sir Isaac Newton*, I, 116.

[11] Each experiment which Newton produces in order to demonstrate the value, and even the *uniqueness* of his theory (cf. n.7 for a brief outline of this feature of Newton's methodology) is to be regarded as a more or less effective 'realization of an elementary process' in the sense of Hugo Dingler (see the latter's *Methode der Physik* (Munich, 1938), esp. 146 where the realization of the 'elementary process' of gravitation is discussed. Newton himself has described the manner in which 'the elementary process of colour separation' might be realized and has indicated the obstacles lying in the way of such realization: *Phil. Trans. R. Soc. Lond.* A, no. 80 (1671), 3087). Dingler's investigations, which unfortunately are not very popular today, deserve detailed study especially in connection with Galileo's and Newton's use of experiment. Cf. also his interesting and challenging book *Das Experiment* (Munich, 1928).

That the experiments of Galileo are illustrations rather than evidence has been shown most convincingly by the late professor N. R. Hanson in 'Galileo's Discoveries in Dynamics' *Science*, 167 (1965), 471–8.

That the same is true of Newton's experiments becomes evident from the fact, to be mentioned presently, that Newton did not ascribe equal importance to all experiments involving colour, no matter how carefully they were devised. 'As others had tailored the experiment to their needs, so also Newton tailored it to his' (Westfall, *Isis*, 53 (1962), 351). This becomes especially clear from his reply to Lucas, quoted below. Lucas had provided what Newton had invited everyone to do, viz. 'other experiments which directly contradict[ed him] (Queries, Cohen, *Newton's Papers and Letters*, 94). He was thanked for that as the 'first who has sent me an experimental examination' (reply to Lucas, Cohen, 173). But he was also advised to 'change a little the method which he has propounded, and instead of a multitude of things try only the *experimentum crucis*. For it is not the number of [in the present case, different] experiments but weight to be regarded; and where one will do, what need many? . . . The main thing he goes about to examine is, *the different refrangibility* of Light. And this I demonstrated in the *experimentum crucis*. Now if this demonstration be good, there needs no further examination of the thing . . . Let that experiment therefore be examined in the first place.' (Cohen, 174). The point of the quotation is that there are some experiments which clearly exhibit the basic principles under dispute whereas others are only remotely connected with them. It is interesting to consider that Newton seems to have been led to this classification of experiments by the very same mechanical philosophy he tried to replace later by his own point of view (Westfall, *Isis*, 53 (1962), 351).

experimentum crucis. 'Now if this demonstration be good, *there needs no further examination of the thing* . . . and seeing I am well assured of the truth and exactness of my own observations, I shall be unwilling to be diverted by any other experiment from having a fair end of this made in the first place.'[12] Phenomena, then, are *selected* and *idealized* experiments whose features correspond point for point to the peculiarities of the theory to be proved.[13]

[12] Cohen, *Newton's Papers and Letters*, 174.

[13] It follows that phenomena have the logical status of *laws*. This conjecture is confirmed by a look at the *Principia* where all of Kepler's laws are found among the phenomena. And the optical 'phenomena' too do not merely express what occurs in a single experiment, but what emerges in all experiments of a certain kind. This feature enables Newton to make good his promise (cf. n.7) to derive (other) laws from them. For while the derivation of laws from singular statements is a problematic affair (Hume's problem!) the derivation of laws from general statements is not. (Hume's arguments are therefore irrelevant for a criticism of Newton.)

Now what we have said so far, while essentially correct, gives only a rough account of Newton's procedure. The next step would consist in pointing out that not only theories but also the so-called phenomena themselves are 'made liable to exceptions' (cf. n.7, rule 4 as well as the formulation of this rule in the text to n.18).

These exceptions are mentioned in the very beginning. Thus in stating phenomenon 6 of book 3 of the *Principia*: 'that the moon, by a radius drawn to the earth's center, describes an area proportional to the time of description' Newton comments 'It is true that the motion of the moon is a little disturbed by the actions of the sun; but in laying down these phenomena I shall neglect those small and inconsiderable errors' (*Principia*, 405). For this and other references to Newton's *Principia*, see *Mathematical Principles of Natural Philosophy*, trans. A. Motte, ed. and rev. F. Cajori (Berkeley, 1960). Phenomenon 5 reads that 'the areas which the primary planets describe by radii drawn to the sun are proportional to the times of description' (*Principia*, 405) while a few pages later (421) we are informed 'that the action of Jupiter upon Saturn is not to be neglected . . . and hence arises a perturbation of the orbit of Saturn in every conjunction of this planet with Jupiter, that astronomers are puzzled with it'. This separation of *phenomenon* and *actual fact* and the definition of phenomena with the help of theory completely reverses the position expressed in Newton's methodology as outlined in n.7. For the exceptions which are observed are now accounted for by the very same theory that is derived from the phenomena as stated without the exceptions (Hegel seems to have been the only thinker to correctly appreciate this feature of Newton's procedure; cf. his *Encyclopädie der Philosophischen Wissenschaften*, ed. G. Lasson (Leipzig, 1920), 235ff). Used in this way phenomena are now no longer a basis of knowledge but they are *conjectures* put forth in accordance either with some rule of simplicity, or with the theory to be demonstrated (this difference between basic methodology and procedure corresponds to the difference between critical practice and dogmatic ideology outlined in section 1).

We get here a very clear insight into the function of phenomena in actual research. *Originally* (Kepler) they formed theoretical conclusions of an elaborate analysis of observational data. In the *Principia* they form part of an argument leading back and forth between the actual observations, which now already contain the deviations from Kepler's laws, and the new point of view of Newton. However they are arbitrarily separated from the remainder of the argument and given special importance so that a vacuous rule of faith may with their help lend additional support to this point of view. It is a marvellous accident, worthy of an equally marvellous explanation that this way of building up the theory of gravitation did not create a miserable patchwork but a coherent system that astounds by its simplicity and its effectiveness in the mastery of concrete phenomena. We shall see very soon (text below) that light did not submit to this procedure quite as readily. What we obtain here is a system that has been called, and not without justification, 'an incoherent and uncertain theory, a theory so full of contradictions and lacunae that one is surprised to see to what extent it could convince the majority of the physicists of the 18th century' (V. Ronchi, *Histoire de la Lumière* (Paris, 1956) 191).

It is clear that one should rather regard them as illustrations of special consequences of that theory.[14]

8. The second identification is between phenomena and what the empirical rule of faith regards as the foundation of all empirical knowledge, viz. *experience*. We see here again how necessary is the vacuity of the experimental philosophy whose support is enlisted by the two identifications.[15] All that is known about experience is that it is something that springs to the eye, that it is a 'divine illumination' this time not by God, and not through the mind, but by Nature, and through the senses; but which guarantees success. It is therefore again quite easy to turn part of the new theory into its own foundation by first presenting selected phenomena in its terms and by then pronouncing these phenomena, these well-illustrated pieces of the theory to be the experience that has proved the theory 'positively and directly [and not only] from the confutation of contrary suppositions'.[16]

9. Both identifications go almost unnoticed. Hooke, for example, Newton's main opponent, 'regards the phenomena described as facts which leads to much silent advantage for Newton'.[17] And now the development is as follows: those interested in light familiarize themselves with the phenomena, i.e. with the basic principles of the theory. The detailed manner in which Newton develops this theory allows them to apply it to natural processes also. Everything is now being seen in terms of rays being separated, reunited, absorbed, reflected, but never changed in their intrinsic quality. The theory, then, is very *successful*. When we consider the general attitude of the Royal Society it is quite natural to relate this success to the empirical rule of faith: the theory is successful *because* the rule has been obeyed. Conversely, obeying the rule gives results and advances knowledge. Attacks upon the theory are soon answered by pointing out how firmly it rests on experimental fact. Attacks upon empiricism are answered by quoting the successes of the empirical rule of faith such as the theory of colours and the theory of gravitation. A little later one altogether ceases to discuss attacks of this kind, one simply shrugs them off as further examples

[14] This and other features to be given presently make it clear that Newton's method must be described as a *conventionalism* with illustrations. The selective attitude towards experiments also establishes a relation to Einstein who refused to let his appraisal of a theory be influenced by 'verification through little effect[s]' (quoted from G. Holton, 'Influence on Einstein's Early Work' *Organon*, 3 (1966), 242).

[15] Cf. sections 4 and 5.

[16] Queries of 8 July 1672, Cohen, *Newton's Papers and Letters*, 93.

[17] Ipsen, *Farbenlehre*, 614. 'The Society' writes Goethe 'had hardly come into existence as Newton was received into it, in his thirtieth year of age. Yet how he was able to introduce his theory into a circle of men who were most definitely averse to theories, this is an investigation well worthy of a historian.'

of the paradoxes created by armchair philosophers. And so the delicate collaboration between a vague rule of faith and an increasingly popular doctrine has again created a bastion that can withstand the strongest attack.

But Newton is still not content. The bastion must be defended even more strongly by making outside attacks impossible in principle. This is done with the help of the famous *rule 4* of the *Principia*: 'In experimental philosophy we are to look upon propositions inferred by general induction from phenomena as accurate, or very nearly true *notwithstanding any contrary hypothesis that may be imagined* till such time as other phenomena occur by which they may either be made more accurate, or liable to exceptions.'[18] The importance of the italicized clause becomes clear when we now compare the ray theory with the only developed alternative which existed in Newton's life time, *Huyghens' wave theory*.

10. According to Newton, the wave theory is unacceptable because it cannot explain the rectilinear propagation of light: 'For me the fundamental supposition [of the wave theory] in itself seems impossible; namely, that the *waves* or vibrations of a fluid can, like the rays of light, be propagated in straight lines, without a continual and very extraordinary spreading and bending every way into the quiescent medium, when they are terminated by it.'[19] This attitude is retained despite the phenomena of *diffraction* which Newton has described in detail. The rationale that in these cases the 'bending is not towards, but from the shadow' (*Opticks*, query 28) is strange in view of the fact that Newton himself reports a bending into the shadow, in his communication of 1675, quarrelling here with Hooke about the peculiar explanation he proposes (the light is refracted by the ether which surrounds the obstacle in a thin layer). The same quarrel appears in the *Opticks* (book 3, part 1, observation 5). However, the method of letting basic principles be decided not indiscriminately, by every and any experiment, but by paradigmatic cases which visibly demonstrate the truth of the theory helps us over this difficulty. Having established, both by the *experimentum crucis* and by the swift disappearance of fixed stars behind the moon (*Opticks*, query 28) that light consists of *rays* he now accounts for diffraction by additional assumptions, adding 'new original properties . . . besides those already described' (query 25).

This procedure is indistinguishable from the framing of *ad hoc* hypotheses. It is especially helpful in the case of *mirror images* which the ray theory, taken by itself, fails to explain: 'If light were reflected by impinging upon

[18] *Principia*, ed. Cajori, 400.

[19] Reply to Hooke. The waves which Newton uses to explain interference and which some authors have regarded as evidence that he held a wave theory also are never identified with light but are explicitly regarded as *separate entities* which *interact* with light. Cf. the paper of 1675.

the solid parts of the glass it would be scattered as much by the most polished glass as by the roughest' (book 2, part 3, proposition 8; query 31). In the paper of 1675 we read the hypothesis that the ether leaking beyond the roughly aligned atoms of the mirror provides a smooth surface, which fits ill the aversion to hypotheses shown on other occasions (cf. n.7). In the *Opticks* 'the reflection of a ray is effected, not by a single point of the body, but by some power . . . which is evenly diffused all over the surface and by which it acts upon the ray without immediate contact' (cf. above), which is just a description ('soporific power'!) of the *fact* of mirror images in terms of the ray theory. We see here very clearly how Newton preserves basic assumptions that readily lend themselves to illustrative demonstration with the help of 'new original properties', i.e. *ad hoc* hypotheses.

A method like this, which starts from a paradigmatic case and adds new assumptions as soon as it appears that the theory expressing the paradigm is in difficulty, will not easily discover 'other phenomena by which [the theory] . . . may be made . . . liable to exceptions' (rule 4), *unless* one is allowed to do what rule 4 explictly forbids, viz. to 'imagine contrary hypotheses' such as, for example, the wave theory.

11. A simple look at this theory suffices to replace the success story of the ray theory by a very different tale. The wave theory, too, has its paradigmatic experiments which exhibit its principles at once, and with only a minimum of abstraction and generalization: the phenomena of refraction and reflection follow from this theory as swiftly and as naturally as the *experimentum crucis* follows from Newton's account. Applying the method of modifications by addition, suggested by Newton, we can explain rectilinear propagation in a way that involves no new idea and is much more satisfactory than is Newton's account of mirror images, and diffraction: the partial waves which spread outside the cone of propagation 'do not concur at the same time to compose a wave'; these parts are therefore 'too feeble to produce light'; 'whence one sees the reason why light . . . spreads only in straight lines, so that it illuminates no object except when the path from its source to that object is open along such lines.'[20] Polarization needs no 'new original properties' but is a feature of the basic (transversal) process itself. Moreover, this basic process also suffices for explaining diffraction. It therefore already contains three fundamental properties of light (reflection; refraction; polarization) and provides means for calculating the others. Looking back from this success (which occurred much later, in the nineteenth century) to the original paradigm (mirror images) as well as to some more recondite phenomena such as diffraction we now realize that these phenomena not merely add 'new original properties' to the ray theory but that they also refute it while the rectilinear propagation of light, if taken

[20] Huyghen, *Treatise on light*, transl. S. Thompson (New York, 1962), 16.

together with diffraction, *lends support* to the wave theory. This story is very different from Newton's claim to have finally established the basic assumption of the ray theory. (Note that we do not proceed here beyond the facts which were known to Newton himself.)

12. Now it is exactly this kind of criticism that is excluded by the italicized clause of rule 4. The criticism can be articulated only if we are allowed to view the success of Newton's theory in the light of 'contrary hypotheses', and if we are permitted to elaborate and entertain such hypotheses despite their *prima facie* implausibility. If on the other hand we follow rule 4 to the letter, then contrary hypotheses will not be used and the criticism cannot arise.[21] Quite the contrary: one will be able to point to the magnificent success of the theories which have been developed with the help of the rule. One will be able to point out that rule 4 does not lead to stagnation. The scientists who adopt it heap discovery upon discovery and continually expand the domain of knowledge. This is a strong argument in favour of the rule and in favour of all the theories supported by it.

This argument, of course, cannot really satisfy us. For just as in the case of Protestantism the success of the chosen theories is entirely man-made. It is due to the fact that the psychological result of a complex process of indoctrination was turned into a basis. In the case of Protestantism the basis supported a *faith*. Here it supports a scientific *theory* which is constantly being expanded by the addition of *ad hoc* hypotheses (this is what the 'success' of the theory really amounts to). In both cases we are dealing with nothing but a party line.

13. Let us reconsider what we have discovered so far. We have discovered that the rejection of authority, of tradition, of the results of speculation that is such a characteristic outer feature both of Protestantism and of the empiricism of Bacon does not lead to a more critical attitude. It leads to the enthroning of new authorities which demand slavish attention: scripture on the one side, experience on the other. We have also discovered that the rules of faith which introduce the new authorities are vacuous. We have seen how the very vacuity of the rules makes them excellent allies in the defence of partisan ideas. If we follow the demand for an authoritative foundation such ideas are first made plausible, they are then based upon their own most plausible parts and are justified thereby. In the case of Protestantism the plausible parts are the intuitions resulting from a strict and merciless education. In the case of empiricism the plausible parts are those elements of a theory which can be readily illustrated by experiment. The physical stability of those experiments is then seen as proof of the stability of the experience recommended by the rule of faith. This is the main feature of the

[21] Rule 4 therefore corresponds to condition (2), in section 3.

classical empiricism that forms the subject of the present chapter. But illustrations are not evidence. Intuition is no objective guarantee. Physical stability of a piece of matter must not be confounded with epistemological certainty. A method that first makes an idea familiar, either by frequent repetition, or by illustration, and that later on uses this plausibility as if it were an additional source of support is not different from political propaganda, and ideas thus defended are, as we have said, indistinguishable from party lines.

14. Now this situation, while apparently quite deplorable, in fact constitutes a tremendous step forward from the preceding philosophical views. For while the vacuity of the new rules of faith allows a circular defence of ideas which have achieved a certain notoriety, it also guarantees that no particular idea is preferred. *Any* idea can now be presented in a manner that makes it acceptable and capable of winning followers. Of course, there will always be some views which are in the foreground and try to demonstrate their uniqueness, while others are not yet sufficiently developed and sufficiently familiar to attract attention. This, however, is only a psychological disadvantage. Given time, colourful and persistent defenders, surprising successes or sham-successes, and satisfaction of special interests, the support of the rules of faith may be enlisted for a quite different philosophy. This very neutrality shows that they are but an ornament, surrounding whatever convictions we possess and presenting these convictions, our own fallible products, in an objective manner, as if they had their origin elsewhere and derived their importance from an independent and not human source. The fact that we are dealing with party lines is therefore not really a drawback. Quite the contrary: party lines play a most important role in many civilized institutions, such as in the democratic process, in the process of trial by a competition of opposing opinions which allows for the examination of even the most fundamental assumption and the most expert testimony, and so on. In the last case the creation of additional party lines, of 'contrary hypotheses' is even *demanded* by the law as a means of seeing in the proper perspective a theory or some expert testimony which, *taken by itself*, would seem to be totally invincible (remember the ray theory!) Party lines are not the problem. Problems arise only when an attempt is made to turn the *subjective* conviction that makes a particular party line stand out into an infallible *objective* judge who withstands criticism and demands that his dictum be obeyed. Classical empiricism which adopts this procedure has not yet completely overcome its even more restricted ancestry. But the democratic way in which praise, blame, and dogmaticism are now distributed and the humanitarian way in which the word of a clever man is taken seriously, even *too* seriously, allow us to greet it as the dawn of an even more enlightened future.

3
The structure of science

1. Science, says Professor Nagel,[1] 'takes its ultimate point of departure from problems suggested by observing things and events encountered in actual experience' (78); 'it aims to understand these observable things by discovering some systematic order in them' i.e. by explaining them (78). It is this 'desire for explanations which are at once systematic and controlled by factual evidence that generates science' (4; see also 15). Understanding the structure of science, therefore, means understanding the structure of scientific explanations (15).

The view of explanation set forth by Nagel does not differ in any essential respect from the hypothetico–deductive account. However, a considerable amount of detail is added. In what follows I shall deal with the distinction drawn between experimental laws and theories (ch. 5); the cognitive status of theories (vi); the explanation of established theories (laws) by other theories (ch. 11); and with Nagel's general views concerning the structure of science.

2. Theories are distinguished from experimental laws because they 'employ terms like "molecule" which ostensibly designate nothing observable' (80). The distinction does not mean that laws are more highly confirmed than theories, or that theories are 'entirely speculative' (80). Nor is it assumed that experimental laws express 'relations between data . . . apprehended directly or non-inferentially through various sense organs' (81; 122). The vagueness of the term 'observable' (83; 90); the difficulty of proposing a 'precise criterion' distinguishing theories and experimental laws are duly acknowledged. Yet the inference that 'the distinction is spurious because it is vague' is rejected by asserting 'several well marked features' differentiating the former from the latter (83). These 'several features' essentially boil down to one: each descriptive term in an experimental law 'is associated with at least one overt procedure for predicating [it] to some observationally identifiable trait when certain specified circumstances are realised' (83). This feature connects with others, viz. (i) the meaning of the descriptive terms of experimental laws, the so-called observational terms, is partly 'fixed' by the procedure; hence, (ii) 'an experimental law, unlike a theor-

[1] Ernest Nagel, *The Structure of Science* (London, 1961).

etical statement, invariably possesses a determinate *empirical content*' (83); (iii) experimental laws but not theories can be the result of inductive generalizations (85); (iv) to the extent to which they are connected with operational procedures the content of experimental law *is* independent of any theory that might be used for their explanation. It also *must* be independent, 'the law must be intelligible (and must be capable of being established) without reference to the meanings associated with it because of its being explained by (some) theory. Indeed, were this not the case for the laws which a given theory purportedly explains, there would be nothing for the theory to explain' (87). Hence, 'an experimental law has . . . a life of its own, not contingent on the continued life of any particular theory that may explain the law' (87).

Theories are analysed into (*a*) an abstract calculus whose postulates 'assert nothing since they are statement forms rather than statements' (91) but 'implicitly define the basic notions of the system' (90; 160); (*b*) correspondence rules, relating theoretical notions to 'observational procedures' (94) or 'experimental concepts' (95). Such rules make it possible to use the theory 'as an instrument of explanation and prediction' (93; 106), and to investigate 'its material truth or falsity' (222). Without correspondence rules a theory is not even a statement (141) as its descriptive terms (or rather those for which no rules of correspondence are given) have the status of variables (132; 95). Finally, (*c*) a theory contains a model. A model is defined relative to the abstract calculus and consists in a specification of its postulates that turns them into true statements and preserves the logical relations between them (96, n.4). A model in this sense need not provide correspondence rules (95, 97) although it can suggest 'at what point [such] rules may be introduced' (113). Nor does a model turn the postulates into statements and the theoretical terms into constants (132). For while it might contain constants only, it is not the only possible model (97). This also explains why even overwhelming evidence for a theory 'may not be assessed as sufficient for asserting the physical existence of various elements in the substantive model in terms of which the theory is formulated' (117). The calculus, the correspondence rules, the models must 'not . . . be construed as separate items, introduced in succession at various stages in the actual construction of theories, but simply as features that can be isolated for purposes of analysis; (106f).

3. Now it appears that these features can also be isolated in the case of experimental laws and, more generally, in the case of any set of observation statements that possesses suitable closure properties. *Observational terms* are 'associated with overt procedures for predicating them of observationally identifiable traits' or with *ostensive rules* as we shall call them.[2] *Theoretical*

[2] I have taken this very apt term from S. Körner *Conceptual Thinking* (New York, 1962), 7.

terms are related by correspondence rules to observation terms which in turn are related to observable traits by ostensive rules. Separating the theoretical terms from the correspondence rules leaves an uninterpreted system which can be expressed in the form of a calculus C_T with postulates P_T (91). Separating the observation terms from the ostensive rules leaves again an uninterpreted system which can also be expressed in the form of a calculus C_O with postulates P_O. The objection that the 'logic' of the observation terms is completely determined by the ostensive rules cannot be raised by Nagel who rejects sense data (88, 122), who asserts quite explicitly that ostensive rules define a 'partial meaning' only (88) and who points out that already commonsense 'involves the use of abstract conceptions' (11). The further objection that the logical structure of the observation language is hardly ever available in the form of explicit principles overlooks that the same is true of theories also. Here too, the explicitly formulated laws mirror only part of the 'logical relations in which the terms stand to one another'; the remainder must be unearthed by attending to the way in which theoretical terms are being used. It is plausible to assume that many theoretical postulates have originally been obtained in this manner (examples: the Archimedian postulate in geometry; the law of excluded middle). In the case of observation languages this applied to *all* postulates – this is the only difference. Finally, there will be *models* for P_O. The most familiar models of course assume things having certain properties, but we need not restrict ourselves to them. We may use models similar to those introduced by Whitehead in the case of geometry, and even more unfamiliar ones. There emerges therefore a complete parallelism, in all the respects examined by Professor Nagel, between the domain of observation and the domain of theories. Let us examine the consequences of this parallelism.

4. It is often said that the aim of science is the classification and the prediction of observational results, and that the distinctive characteristic of *modern* science is mathematics and the experimental method. This is correct as a first approximation, but one has to add that experiments are used not only for discovering new facts, but also for revealing the detailed structure of facts already known. This is what really distinguishes modern science from its predecessor. The laws and descriptions of the Aristotelian science were supposed to give an account of observationally identifiable traits *exactly as they appeared to the observer* (cf. the famous example of the drop of wine in 10,000 pails of water, *de gen. et corr.*, 328a27). The point of view of Galileo and Newton (which was prepared by the ancient atomists) aims at an *analysis* of such traits that first eliminates the contribution of the observer and then tries to show how the remainder results from an interplay of processes which can be isolated only with difficulty, or perhaps not at all (and which will have to be described by theories in Nagel's account). We

see already here that the familiar characterization of science as an enterprise that 'aims to understand . . . observable things by discovering a systematic order in them' (Nagel, 78) is still quite Aristotelian, especially if we take it together with the remark that observational laws and observational notions 'have a life of [their] own' (87). And we anticipate that a different account will have to be given of the role of observation and experiment.

A science that is interested in analysis uses theories for describing the 'hidden' processes which constitute the observable traits. Such theories must be rich enough to express the specific features of the *individual* processes, and they must also be capable of giving an account of the manner in which all these processes *collaborate* and bring about the overt traits and observational facts to be analysed. In other words, they must provide statements and laws suitable for expressing such facts. Of course, the statements need not *repeat* what was said about the facts at an earlier time. Nor is it advisable that they should. The earlier formulations will usually contain the gross reactions of the observer and the crude beliefs emerging from them and it is just such beliefs one wants to exclude. All that is needed is that the order introduced into the observable material by the theory be comparable, both in complexity and in effectiveness, with the order established by the concepts and laws already in use. The exact form of the order need not be preserved. Thus the explanation of Kepler's first law by the general theory of relativity (Schwarzchild's solution) does not give us an exact ellipse, nor are the predicates describing the shape those of ordinary physical geometry. They do not even share a core of meaning with the preceding notions.[3] Yet the descriptive apparatus of the explanation can compete well with that of the older astronomy. It is even preferable as it no longer contains the incorrect assumption of a spatiotemporal structure that is independent of the state of motion of the observer. It follows, then, that a satisfactory theory contains a calculus C_T^O and postulates P_T^O which, though not coinciding with C_O and P_O can yet provide just as effective a logical skeleton for an observation language.

Let us return to Professor Nagel's account. In this account the ostensive rules are applied to the descriptive terms of C_O; the terms of C_T, C_T^O included, get their interpretation only indirectly, with the help of correspondence rules. What has just been said would seem to suggest that C_T should perhaps be interpreted in a more direct fashion, without the detour through C_O and correspondence rules. There is no reason why the overt procedures which according to Nagel are used to 'fix a definite, if only partial, meaning' for the descriptive terms of C_O should not be applied to the descriptive terms of C_T^O and in this way fix a definite, if only partial, meaning for *them*. The remainder of the theory would receive its content not

[3] See vol. 1, ch. 5.

through additional linkages, added from without, but via the deductive relations between P_T, C_T and P_T^O, C_T^O. C_O, P_O and correspondence rules would be superfluous. For example, one might apply the operations usually employed for giving content to the term 'heavier than' of some commonly understood idiom to the term 'possessing much greater relativistic mass than' of the special, or the general theory of relativity.

This *direct method of interpretation* as we shall call it would seem to reflect much better the features of a science that progresses by analysis than does Nagel's method. It avoids unnecessary duplication of terms and relational systems (C_O with P_O versus C_T^O with P_T^O). It prevents the perennial retention of the errors implicit in the older observational idiom (e.g. it prevents the perennial retention of a language that is not Lorentz invariant; or of a language of mental terms that does not allow for the incorporation of materialistic features). The charge of circularity which Nagel might raise and the related charge that such a decisive change of the observation language may leave 'nothing for the theory to explain' (87) counts little in comparison with such advantages. Besides, it adds another point in favour of the direct method. For the need to change the observation language indicates that there was no explanandum to start with, or at least, that the older experimental laws were not the proper explanandum. What *did* need explanation was how people could believe them, but this is an altogether different matter.

5. Nagel's views on the *interpretation* of theories are closely connected with his account of a *reduction* which is 'the explanation of a theory, or a set of experimental laws established in one area of inquiry [the so-called secondary science] by a theory usually, though not invariably formulated for some other domain [the primary science]' (338). The most striking cases of reduction are those where a 'set of distinctive traits of some subject matter is assimilated with what is patently a set of quite dissimilar traits' (339f). In these cases the 'laws of the secondary science' contain terms that do not occur in the theoretical assumptions of the primary discipline' (352). A reduction will now be possible only if 'special hypotheses', connecting the dissimilar terms of both disciplines are used in addition to the principles of the primary science (365; 353f). The *existence* of these hypotheses must not be understood as expressing an 'ontological hiatus' (365) or some 'experimental, or even "metaphysical" facts about some allegedly "inherent" . . . *properties* of objects' (369). It rather expresses the '*logical* fact' that the terms of two sets of statements stand in certain relations to each other (369).

The *content* of the connecting hypotheses depends on the context in which a reduction is to be achieved (357). The principles of the primary science may be connected with experimental notions by these hypotheses only: they 'cannot [then] be checked by experiment, but function as coordinating definitions' (356) (case 1). On the other hand the principles of the primary science may have been interpreted independently of, and perhaps already before, the secondary science. The reduction of thermodynamics to mechanics is an example of this kind. In this case the 'expressions belonging to [both the principles of the primary and the secondary science] possess meanings that are fixed by [their] *own* procedures of explication . . . [and] are intelligible in terms of the rules or habits of usage of the [corresponding] branch of inquiry; and when those expressions are used in that branch of study, they must be understood in the senses associated with them in that branch, whether or not the science has been reduced to some other discipline' (352, a repetition, on a higher level, of the remark about the independent life of experimental laws). Clearly the connecting hypothesis 'will [now] be one that does not hold as a matter of definition, and will not be one for which logical necessity can be claimed' (358) (case 2). Most reductions are supposed to be of this kind. Thus the reduction of thermodynamics to mechanics, or of chemistry to contemporary physical theory 'does not wipe out, or transform into something insubstantial, or "merely apparent" the distinctions and types of behaviour which the secondary discipline recognises', nor does the reduction of headache to neurophysiology 'establish a logically necessary connexion between the occurrences of headaches and the occurrence of certain events and processes specified by physics, chemistry, and physiology' (366). The doctrine of emergence, to mention another case to which the above considerations are applied, is on firm ground 'when construed as a thesis concerning the *logical relations* between [statements describing the emerging traits and statements of the quite different theory used to explain their emergence]' (372). And so is organismic biology insofar as it asserts that connecting hypotheses linking biological terms to terms of physics are necessary for a reduction, but are not yet available (434).

6. From these remarks there now emerges the following view of the structure and, possibly, the development of science.

In so far as they can be related to each other the explanatory systems, or the theories of science, are arranged in *layers*, or *levels*, starting with an observational level and moving up to levels of increasing abstraction. Each level is self-contained. Having methods and procedures of its own (352) it gives rise to an equally self-contained domain of knowledge. The layers are 'linked' by 'rules of correspondence' with the observational level (93ff) whose laws have 'a life of (their) own, not contingent on the continued life of any particular theory that may explain them' (87). In addition the layers may be connected with each other by 'specific hypotheses' (365; 473). The rules of correspondence give empirical content (90, 166, 300) and they also make possible the explanation of observable phenomena (93, 97) which are the 'ultimate points of departure' for 'scientific thought' (79). Connections *between* layers are necessary if lower levels are to be explained by, or reduced to, the higher levels. As we have seen such explanation neither 'wipe[s] out' nor 'transform[s] into something insubstantial' the level explained (366). It leaves intact the self-contained character of each level, and the relative independence of each domain of knowledge.

Now if we take it for granted, as does Nagel, that explanation is essential for science (3); if we assume that it is 'the desire for explanations which generates science' (4); then the above arrangement of layers becomes a model not only of the *end result*, but also of the *development* of our knowledge. The development of knowledge then consists in the accumulation of facts, and of theoretical layers. Science advances by internal improvement of each level, by addition of new facts on the observation level, by addition of new explanatory systems on top. There may of course be a considerable amount of change preceding the establishment of each single layer. Theories may have been formulated tentatively, and may again have been abandoned. But as soon as a theory is 'established in one area of inquiry' (138) as soon as it starts pervading this area with its methods and its specific terminology, in the very same moment it assumes the independence of procedures and rules of usage that we have reported above. The lower levels then actually become somewhat assimilated to the observational language and their abstract conceptions assume a similar 'life of [their] own'. Here, then, we have the outlines of what one might call the *layer model of scientific knowledge*.

7. This ingenious model which tries to combine the empiricist's concern for *facts* with the realization of the importance of *theories* and which also tries to satisfy the wish for some amount of stability is in agreement neither with actual science, nor with reasonable methodological demands. The reasons for this have already been stated although so far they have been restricted to the relation between theories on the one hand and experimental laws and observation languages on the other. It has been pointed out that Nagel's method involves an unnecessary duplication of relational systems. And it was suggested that the ostensive rules which connect terms with overt operations and traits should be linked to the theories directly, and not via an older observation language and correspondence rules. In this way the older observation language would be abandoned and its abstract concepts replaced by the abstract concepts of the theory. The charge of circularity was answered by showing that the non-circular procedure preserves (at least part of) the structure of the observational laws and thereby removes it from the domain of criticism. This point deserves repetition: If it is the business of science to explain; if explanation 'takes its ultimate point of departure from problems suggested by observing things and events in common experience' (79); if the description of these things is allowed to

have 'a life of its own' (87) that can be expressed 'independently of [any] theory' (86); if a change of theory changes only the 'theoretical interpretation' of the experimental laws and observational terms, but leaves unchanged another part of their meaning (87), then the relations between terms which constitute this 'observational core' will forever be exempted from change. Their stability will be guaranteed not because a detailed examination has found them to be adequate (Nagel admits that empirical laws may be as 'speculative' as theories (80)), but because they have been arranged in such a manner that an examination becomes impossible. They have been turned into veritable *idola fori*. And such *idola* can be criticized only by a procedure that permits us to compare them with other relational systems, for example with the observation languages which universal theories give on direct interpretation. Such a comparison, though being perhaps 'circular' is thereby shown to have a definite advantage over and above the method described by Nagel. Now in the case of reduction the situation is exactly the same. But it is somewhat easier to present it in an acceptable manner.

Assume for that purpose that we are dealing with the explanation of classical (Newtonian) momentum conservation by the special theory of relativity. Nagel himself regards the relativistic notion of mass as a 'new notion' and he points out that the conservation principles 'must be reformulated in terms of relativistic mass' (111). He gives a brief, though very enlightening, account of the manner in which the relativistic concept differs from the classical concept (170). Newtonian mass is an 'intrinsic property' and it is 'additive'. Relativistic mass is a function of the relative velocity (and it is non-additive). True, he seems to be inclined to restrict the modification to particles moving with great speed ('the mass of a particle moving with great speed varies with the velocity, so that the [classical] principle of conservation of momentum does not appear to hold for *such* principles' (111, my italics); but I do not think that this passage should be interpreted as saying that the dependence on velocity *disappears* at low speeds (although it is of course to be admitted that it can no longer be *observed* at low speeds). Now the first thing we note is that at low speeds, and after omission of effects unlikely to be spotted experimentally, the theory of relativity gives *exactly the same predictions* as does Newton's theory, and we also obtain a 'sentence *similar in syntactical structure* to the standard formulation [of Newtonian momentum conservation]' (358, my italics; the reference is to the reduction of thermodynamics). In the domain in question ($v/c \ll 1$) the theory of relativity is therefore as effective a predictive device as is the older point of view. But of course, the 'similar syntactical structure' has a meaning that is 'unmistakably different from' what Newtonian momentum conservation asserts (358; the reference is again to thermodynamics). This follows from what was said above: the theory of relativity uses the relativistic notion of mass (dependence on v/c; non-additivity) under *all* circumstances, and not only for velocities comparable with the velocity of light although for $v/c \ll 1$ the predictions (i.e. the *numbers*) resulting from such use will be indistinguishable from the numbers produced with the help of the classical terms. We see then that the theory of relativity can do everything the classical theory can do; that it can do a few more things; and that it is successful where classical physics fails. And it would seem that we can safely *abandon* the classical point of view and live with relativity instead. Of course this does not mean, as has sometimes been suggested, that we shall abandon the splendid mathematical machinery of classical celestial mechanics and that we shall regard the achievements of this discipline as being of historical interest only. Abandoning classical physics means abandoning the classical *concepts*; the classical *formulae* may be retained.[4]

However, this is just the procedure which encounters Nagel's objection. It may be granted by the objection that we have made some progress. But an *explanation* of classical mechanics has not been achieved. Such an explanation demands the continued use of the classical concepts (cf. the argument at the bottom of p. 357, as well as the presentation in section 5). Let us now examine this objection.

The objection is based on the following two assumptions: (i) science proceeds by explanation; (ii) the explanation of well-established laws is by deduction. From these assumptions it is inferred that there must be 'special hypotheses' (cf. section 5) connecting the relativistic notions with the classical notions. Using these hypotheses we obtain pairs of theories whose elements are capable of making predictions in a certain domain, one of them being capable of making predictions in other domains also. The first question that arises is why science should be cluttered up with double systems of this kind when one half can do the job of both. The second question concerns the connecting hypotheses themselves. Here we must distinguish what we earlier called case 1 and case 2 (section 5). Case 1 means now explaining relativistic mass in terms of classical mass, which certainly is not in accordance with Einstein's intentions (though it may be in accordance with the original intentions of Lorentz). Another example will make the situation even clearer: the phlogiston theory, especially in the form in which it was developed by Stahl and the new ideas of calcination and oxidation form a perfect pair for reduction. It does not matter that the phlogiston theory has now been abandoned: quite the contrary, this is an argument in favour of our procedure. What does matter is that at the time in question it was well established and, because of Stahl's particular

[4] Though it seems to me that too much emphasis even on the formulae may prevent us from predicting relativistic effects where we do not expect them.

prejudices, thoroughly observational. The method discussed at the present moment would amount to *retaining* the notion of phlogiston and the theoretical context surrounding it ('the expressions belonging to a science possess meanings that are fixed by its own procedures . . . whether or not the science has been reduced to some other discipline' (352)), interpreting all the new notions by reference to it. It would amount to saying that new conditions for the production and transfer *of phlogiston* had been discovered (cf. the similar locution in the case of headache, 366). However, we prefer to say that the notion of phlogiston had turned out to be inadequate and that we had found reasons to abandon it.

Case 2 assumes that the special hypotheses connecting classical terms and relativistic terms are empirical statements asserting 'that the occurrence of the state of affairs signified by a certain theoretical expression . . . in the primary science is a sufficient (or a necessary and sufficient) condition for the state of affairs designated' by an expression in the secondary science (354). In the present case they would assert that whenever a certain relativistic state of affairs is realized, intrinsic mass that is additive (cf. 170) is present also. But the theory of relativity *roundly denies* the existence of intrinsic masses and therefore contradicts the hypothesis. Again, the method recommended by Nagel leads to undesirable consequences.

Now this method follows from assumptions (i) and (ii) concerning explanation. The fact that it leads to undesirable consequences shows that at least one of the assumptions is in need of change: we shall either have to admit that science does not proceed by explanation, or that explanation (or reduction) is not by deduction. As far as I am concerned the problems as to which way to choose has no interest whatever. If we know how science proceeds, if we know what moves can be recommended as reasonable, then we know all we need to know. The question as to how these reasonable moves should be *described* is a problem for the publicity experts and should perhaps best be settled by introducing a new terminology that is not as loaded as is the term 'to explain'.

8. Our argument is still incomplete. Nagel assumes that the laws of the secondary science have been 'established in some area of inquiry' (338) so that they agree with the experimental results in this area. Now while classical mechanics satisfies this condition (the area of inquiry being defined, relative to special relativity, by $v/c \ll 1$) it is refuted by experiments outside (iii). The same is true of the example which Nagel himself uses, viz. the reduction of thermodynamics to mechanics. Here, too, there exist experimental results (Brownian motion) which contradict the secondary science. Moreover the existence of such common phenomena as noise would seem to show that the domain of success of the secondary science is

really quite small.[5] Now in order to dispel the impression that my argument against Nagel succeeds only because he has chosen his examples badly and that it depends on whether or not the secondary science is adequate I shall assume in the present section that it is and that no exception has been found. Can Nagel's theory of reduction be retained under these quite unrealistically ideal circumstances? I think it cannot. The reasons are as follows.

One of the most important tasks of science is to *test* the laws and the theories it introduces for the purpose of explanation and analysis. This cannot always be achieved by a direct comparison with facts. Some of the most decisive refuting instances for a theory T are produced with the help of *alternative theories* T' which contradict T, repeat the success of T, make predictions in new domains at least some of which are then confirmed by phenomena P. We may say in such cases that T has been indirectly refuted by P. The refutation of thermodynamics by Brownian motion is of exactly this kind. Moreover, it can be shown in this case as well as in others that the alternative T' is *essential*, since a direct refutation of T by P (of thermodynamics by Brownian motion) is excluded by physical laws.[6] Now pairs of theories such as T and T' are ideal candidates for reduction, and this according to Nagel's own criteria. In reducing T to T' Nagel would interpret them in a manner that removes the inconsistency (this follows from his general views on reduction as well as from his own example, the reduction of thermodynamics to mechanics). We have shown above that such interpretations frequently run counter to the intentions of the proponents of T and T'. We have also mentioned other disadvantages. We now see that the procedure lowers the empirical content of our knowledge by eliminating valuable tests. It is therefore consistent neither with actual science, nor with reasonable methodology.

In the meantime I have been informed by Professor Smart that Nagel is prepared to admit the inconsistency of many theories he regards as candidates for reduction and that he suggests construing the connecting hypotheses as probability hypotheses. This does not at all improve things. Firstly, because the remark concerning the unnecessary multiplication of explanatory systems still remains in force. Secondly, the theory of relativity (to mention only one example) excludes intrinsic mass not only with high probability, *but always* (see the argument in section 7, case 2). Thirdly, a weakening of the inconsistency between competing theories would reduce the critical ability of both and must therefore be avoided. No amount of small scale adaptation will do. Nagel's theory of explanation must be altogether rejected.

[5] One should also consider that the existence of a finite specific heat for all objects leads directly to the kinetic theory.

[6] For details see vol. 1, chs. 4 and 6.

9. We now come to the last item on our list, the question of the cognitive status of theories. It is impossible to treat here in detail all the arguments and interesting observations which Nagel presents on this problem. I only want to make two points. First, Nagel points out that 'a defender of either view [realism or instrumentalism] can not only cite eminent authority to support his position; with a little dialectical ingenuity he can usually remove the sting from apparently grave objectives'. And he concludes that 'once both positions are so stated that each can meet the *prima facie* difficulties it faces, the question as to which of them is the "correct position" has only terminological interest' (141); 'in brief, the opposition between these views is a conflict over preferred modes of speech' (152).

Now it is certainly correct that much of the contemporary discussion of the issue has precisely the character described by Nagel. It is also obvious that the addition of further *ad hoc* hypotheses may remove the still remaining differences. But when comparing philosophical theories, or physical theories, we should not use their degenerate forms but those formulations of them which contain only a minimum of adaptive measures. And there are versions of realism and instrumentalism which satisy this demand. The instrumentalist version of the idea of the motion of the earth, to take one of the most important examples, was supported by the argument that a real motion of the earth would contradict the highly confirmed laws of the Aristotelian science of dynamics. The instrumentalistic version of Newton's theory of gravitation was already defended by Newton himself on the grounds that experience had shown matter to be inert so that it could not be regarded as a seat of forces. A realistic interpretation of the quantum theory is rejected on the grounds that it is inconsistent with the conservation laws and the laws of interference. These are strong reasons and they can be removed only by arguments showing that it is desirable to introduce theories which contradict already existing laws. In the last section we have indicated that such arguments can be provided. This removes one of the most serious historical objections to realism.

The second point I want to make is connected with the usual debates concerning the distinction between observation terms and theoretical terms. As has been reported in section 2, Nagel objects to regarding the distinction as spurious because it is vague (83). This is certainly a valid point unless the *raison d'être* of the distinction consists in the role it plays in arguments about matters which do *not* admit of gradual transitions. In this case proof of the vagueness of the distinction is proof of its irrelevance. Now one of the most important reasons for making the distinction (and *the* reason for a thinker like Berkeley) is that one wants to know what is *established*, and what therefore can be *known* tc exist, and what is hypothetical only. We take it for granted that observable things can exist whereas some would claim that ascribing existence to theoretical entities is not only false, but a logical blunder. Now the distinction between existence and non-existence certainly is not a gradual one. Admitting that observability is a vague concept is therefore tantamount to admitting that the problem as to what exists and what does not is independent of it, that it is a metaphysical problem, and that the question as to what can be observed and what cannot be observed is comparatively unimportant. Exactly the same result is obtained by attending to what has been said in section 3.

In this section we pointed out that observation languages which are not sense-data languages, which contain abstract notions, and whose use amounts to asserting certain relations can be dealt with along exactly the same lines as theories. There will be a calculus, C_O with postulates P_O and there will also be models. It follows, then, that even the greatest success of the observation language 'may not be assessed as sufficient for asserting the physical existence of the various elements in the substantive model in terms of which the [observation language] is formulated' (117, applied to theories). Not even the existence of observable events is therefore assured and the 'problem of the cognitive status of theories' expands into the problem of the cognitive status of *all* notions of our language. Nagel's book contains hardly a hint concerning this more general problem and this despite the fact that his own analysis of theories and his rejection of sense data leads straight to it. This is true of other accounts also: the ontological status of observation terms is a matter that is hardly examined today.[7] This certainly is a step back from the sophistication that had already been reached by Kant and an indication that the belief in sense data, though denied in *words* still seems to have a strong influence upon the *deeds* of almost all contemporary philosophers of science.

10. The topics dealt with so far fill hardly one-third of Nagel's book. There are chapters on the quantum theory, on the science of mechanics, and there is a large amount of argument dealing with the social sciences. These arguments are very instructive and I am sure that everyone reading the book will greatly profit from examining them. The reason why I have concentrated on the chapters of the book dealing with more general matters is that the opinions expressed there are widespread, that they are presented in great detail, and supported with better arguments than I have found elsewhere.

[7] An exception is Carnap's article 'Empiricism, Semantics, and Ontology', in *Semantics and the Philosophy of Language*, ed. L. Linsky (Chicago, 1952).

4

Two models of epistemic change: Mill and Hegel

The idea that a pluralistic methodology is necessary both for the advancement of knowledge and for the development of our individuality has been discussed by J. S. Mill in his admirable essay *On Liberty* (quoted from M. Cohen (ed.), *The philosophy of John Stuart Mill* (New York, 1961)). This essay, according to Mill, is 'a kind of philosophical text book of a single truth, which the changes progressively taking place in modern society tend to bring out into ever stronger relief: the importance, to man and society, of a large variety in types of character, and of giving full freedom to human nature to expand itself in innumerable and conflicting directions.'[1] Such variety is necessary both for the production of 'well-developed human beings' (Cohen, *Philosophy of J. S. Mill*, 258) and for the improvement of *civilization*. 'What has made the European family of nations an improving, instead of a stationary, portion of mankind? Not any superior excellence in them, which, when it exists, exists as the effect, not as the cause, but their remarkable diversity of character and culture. Individuals, classes, nations have been extremely unlike one another: they have struck out a great variety of paths, each leading to something valuable; and although at every period those who traveled in different paths have been intolerant of one another, and each would have thought it an excellent thing if all the rest would have been compelled to travel his road, their attempts to thwart each other's development have rarely had any permanent success, and each has in time endured to receive the good which the others have offered. Europe is, in my judgment, wholly indebted to this plurality of paths for its progressive and many-sided development' (268–9).[2] The benefit to the individual derives from the fact that '[t]he human faculties of perception, judgment, discriminative feeling, mental activity, and even moral preference are exercised only in making a choice . . . [t]he mental and

[1] *Autobiography* (London, 1963), 215. Many people are inclined to call Mill a liberal and to dismiss him because of weaknesses of the liberal creed they think they have perceived. This is somewhat unjust, for Mill is very different indeed from much that is called 'liberalism' today. He is a *radical* in many ways. Even as a radical, however, he excels by his rationality and his humanity. Cf. R. Lichtman, 'The Façade of Equality in Liberal Democratic Theory', *Inquiry*, 12 (1969), 170–208.

[2] For one particular element of this plurality, see K. R. Popper, 'Back to the Presocratics', *Conjectures and Refutations* (New York, 1962), 136.

moral, like the muscular, powers are improved only by being used. The faculties are called into no exercise by doing a thing merely because others do it, no more than by believing a thing only because others believe it' (252). Choice presupposed alternatives to choose from; it presupposes a society which contains and encourages 'different opinions' (249), a 'negative' logic (236f), 'antagonistic modes of thought,'[3] as well as 'different experiments of living' (249), so that the 'worth of different modes of life is proved not just in the imagination, but practically' (250).[4] '[U]nity of opinion,' however, 'unless resulting from the fullest and freest comparison of opposite opinions, is not desirable, and diversity not an evil, but a good . . .' (249).

[3] 'Coleridge' in Cohen, *Philosophy of J. S. Mill*, 62. '. . . I had to learn that I would recognize the value of health even in sickness, the value of rest through exertion, the spiritual through deprivation of material things . . . through evil the value of good . . . I suppose all that I ever tried to teach is expressed in these words.' Sybil Leek, *Diary of a Witch* (New York, 1969), 49, 122.

[4] Cf. also my essay 'Outline of a Pluralistic Theory of Knowledge and Action' in *Planning for Diversity and Choice*, ed. S. Anderson (Cambridge, Mass., 1968), which establishes a connection with scientific method.

For the relation between idea and action see *AM*, ch. 2. Emphasis on action within a libertarian framework plays an important role in D. Cohn-Bendit, *Obsolete Communism*, esp. ch. 5, 254: 'Every small action committee [in the customary political language of the West: every institution, however small], no less than every mass movement [every large institution, including government bodies, etc.] which seeks to improve the lives of all men must resolve: (i) to respect and guarantee the plurality and diversity of political currents [in the widest sense of including scientific theories and other ideologies] . . . It must accordingly grant minority groups [such as witches, to mention only one example] the right of independent action – only if the plurality of ideas is allowed *to express itself in social practice* does this idea have any real meaning.' In addition Cohn-Bendit demands *flexibility* and a *democratic base* for all institutions: 'all delegates are accountable to, and subject to immediate recall by those who have elected them . . .' For example, one must 'oppose the introduction of specialists and specialization' and one must 'struggle against the formation of any kind of hierarchy' including the hierarchies of our educational institutions, universities, schools of technology, and so on. As regards knowledge the task is 'to ensure a continuous exchange of ideas, and to oppose any control of information and knowledge.' It seems to me that the best starting point in our attempt to remove the still existing fetters to thought and action is a combination of Mill's general ideas and of a practical anarchism such as that of Cohn-Bendit. Such a combination produces an ideology and a people that refuses to be intimidated, or restricted, by specialist knowledge (including the specialist knowledge disseminated by our contemporary critical rationalists), that tries to reform the corresponding institutions, especially those graceless safe-deposit boxes of wisdom, our universities, and that encourages the free flow of individuals from position to position ('[N]o function must be allowed to petrify or become fixed . . . the commander of yesterday can become a subordinate tomorrow': Bakunin, quoted from James Joll, *The Anarchists*, (London, 1964), 109), assuring at the same time *that every position in society is equally attractive, and is treated with equal respect.* Let no one say that science, being purely theoretical, has nothing to do with action and politics. The scientist whose results are received with respect and even with fear by the rest of the community and whose 'methods' are eagerly imitated lives in a peculiar and often quite constipated environment. It has its own style, its own rules, its own silly jokes, its own standards of 'integrity' which are likely to poison the whole republic unless special preventive measures (elimination of specialists from positions of power; careful supervision of the educational process so that personal or group idiosyncrasies do not become a national malaise; and absolute distrust of expert testimony and of expert morality) are taken. The connection between theory and politics must *always* be considered.

This is how proliferation is introduced by Mill. It is *not* the result of a detailed epistemological analysis, or, what would be worse, of a linguistic examination of the usage of such words as 'to know' and 'to have evidence for'. Nor is proliferation proposed as a solution to *espitemological problems* such as Hume's problem, or the problem of the testability of general statements. (The idea that *experience* might be a basis for our knowledge is at once removed by the remark that '[t]here must be discussion to show how experience is to be interpreted', 208.) Proliferation is introduced as the solution to a problem of *life*: how can we achieve full consciousness; how can we learn what we are capable of doing; how can we increase our freedom so that we are able to decide, rather than adopt by habit, the manner in which we want to use our talents? Considerations like these were common at a time when the connection between truth and self-expression was still regarded as a problem, and when even the arts were supposed not just to please, but to elevate and to instruct.[5] Today the only question is how *science* can improve its *own* resources, no matter what the human effect of its methods and of its results. For Mill the connection still exists. Scientific method is part of a general thory of man. It receives its rules from this theory and is built up in accordance with our ideas of a worthwhile human existence.

In addition, pluralism is supposed to lead to the *truth*: '. . . the peculiar evil of silencing the expression of an opinion is that it is robbing the human race, posterity as well as the existing generation – those who dissent from the opinion, still more than those who hold it. If the opinion is right, they are deprived of the opportunity of exchanging error for truth; if wrong, they lose, what is almost as great a benefit, the clearer perception and livelier impression of truth produced by its collision with error' (205).[6] 'The beliefs

[5] For the propagandistic function of medieval art, see Rosario Assunto, *Die Theorie des Schönen im Mittelalter* (Cologne, 1963), esp. 21–2.

[6] 'Ideological struggle,' says Mao Tse-Tung ('On the Correct Handling of Contradictions among the People', quoted from *Four Essays on Philosophy*, (Peking, 1966), 116), 'is not like other forms of struggle. The only method to be used in this struggle is that of painstaking reasoning and not crude coercion.' '. . . the growth of new things may be hindered in the absence of deliberate suppression simply through lack of discernment. It is therefore necessary to be careful about questions of right and wrong in the arts and sciences, to encourage free discussion and avoid hasty conclusions. We believe that such an attitude can help to ensure a relatively smooth development of the arts and sciences' (114). 'People may ask, since Marxism is accepted as the guiding ideology by the majority of the people in our country, can it be criticised? Certainly it can . . . Marxists should not be afraid of criticism from any quarter. Quite the contrary, they need to temper and develop themselves and win new positions in the teeth of criticism and in the storm and stress of struggle . . . What should our policy be towards non-Marxist ideas? . . . Will it do to ban such ideas and deny them any opportunity for expression? Certainly not. It is not only futile but very harmful to use summary methods in dealing with ideological questions among the people . . . You may ban the expression of wrong ideas, but the ideas will still be there. On the other hand, if correct ideas are pampered in hothouses without being exposed to the elements or immunized from disease, they will not win out against erroneous ones. Therefore, it is only by employing the

which we have most warrant for have no safeguard to rest on but a standing invitation to the whole to prove them unfounded' (209). If 'with every opportunity for contesting it [a certain opinion, or a hypothesis] has not been refuted' (207), then we can regard it as better than another opinion that has 'not gone through a similar process' (208).[7] 'If even the Newtonian

method of discussion, criticism and reasoning that we can really foster correct ideas and overcome wrong ones, and that we can really settle issues' (111–18). The similarity to Mill, whom Mao read in his youth, is remarkable.

It is to be noted that this advice is not put forth generally, but 'in the light of China's specific conditions, on the basis of the recognition that various kinds of contradictions still exist in socialist society, and in response to the country's urgent need to speed up its economic and cultural development' (113; see also 69, i.e. 'On contradiction': '. . . we must make a concrete study of the circumstances of each specific struggle of opposites, and should not arbitrarily apply the formula to everything. Contradiction and struggle are universal and absolute, but the methods of resolving contradictions, that is, the forms of struggle, differ according to the differences in the nature of the contradictions'). Cf. also n.47.

Nor is freedom of discussion granted to everyone: 'As far as unmistakable counter-revolutionaries and saboteurs of the socialist cause are concerned, the matter is easy: we simply deprive them of their freedom of speech' (117. Cf. H. Marcuse, 'Repressive Tolerance' in *A Critique of Pure Tolerance* ed. R. P. Wolff, B. Moore, Jr & H. Marcuse (Boston, 1965), 100. Marcuse's case is quite interesting. He demands that certain powerful elements be excluded from the democratic debate. This assumes that he has the *power* to suppress them and to prevent them from speaking out and making themselves heard. Now, if he has *this* power, then he certainly has the power to make his own views better known, and he has also the power to educate people to think for themselves. One wonders why he prefers to use an imaginary power which he does not yet possess but which he (or his wife) would certainly like to have, for *suppressing opponents* rather than for *education* and a more balanced discussion of views. Does he perhaps realize that well-educated people would never follow him no matter how omnipresent his slogans and how seductive his presentation?)

The restriction occurs already in Mill, though with different reasons, and expressed in different terminology: 'It is, perhaps, hardly necessary to say that this doctrine is meant to apply only to human beings in the maturity of their faculties . . . The early difficulties in the way of spontaneous progress are so great that there is seldom any choice of means for overcoming them; and a ruler full of the spirit of improvement is warranted in the use of any expedients that will attain an end perhaps otherwise unattainable. Despotism is a legitimate mode of government in dealing with barbarians, provided the end be their improvement and the means justified by actually effecting that end. Liberty, as a principle, has no application to any state of things anterior to the time when mankind have become capable of being improved by free and equal discussion. . . .' *On Liberty*, 197–8; cf. Lenin, *'Left Wing' Communism: An Infantile Disease* (Peking, 1970), 40: 'We can (and must) begin to build socialism not with imaginary human material . . . but with the human material bequeathed to us . . .' The difference between Mill and Popper, however, seems to lie in this. For Mill the (material and spiritual) welfare of the individual, the full development of his capabilities, is the primary aim. The fact that the methods used for achieving this aim also yield a scientific philosophy, a book of rules concerning the 'search for the truth', is a side effect, though a pleasant one. For Popper the search for the truth seems to be much more important and it seems occasionally to even outrank the interests of the individual. In this issue my sympathies are firmly with Mill.

[7] This and similar remarks make it clear that Mill (and Popper, who follows Mill in all the respects so far enumerated) is not 'dedicated to a national religion of skepticism, to the suspension of judgement' and that he does not 'den[y] the existence . . . not only of a public truth, but of any truth whatever', as we can read in Willmore Kendall's bombastic but uninformed essay 'The "Open Society" and Its Fallacies', *Am. Pol. Sci. Rev.*, 54 (1960), 972ff, quoted from P. Radcliff (ed.), *Limits of Liberty* (Belmont, Calif., 1966), 38 and 32. To refute the charge of suspension of judgement we should also consider this passage: 'No wise man ever acquired his wisdom in any mode but this; nor is it in the nature of human intellect to

philosophy were not permitted to be questioned, mankind could not feel as complete assurance of its truth as they now do' (209). 'So essential is this discipline to a real understanding of moral and human subjects [as well as of natural philosophy, 208] that, if opponents of all-important truths do not exist, it is indispensable to imagine them and to supply them with the strongest arguments which the most skillful devil's advocate can conjure up' (228). There is no harm if such opponents produce positions which sound absurd and eccentric: 'Precisely because the tyranny of opinion is such as to make eccentricity a reproach, it is desirable, in order to break

become wise in any other manner. The steady habit of correcting and completing his own opinion by collating it with those of others, *so far from causing doubt and hesitation* in carrying it into practice, is the only stable foundation for a just reliance on it; for, being cognizant of all that can, at least obviously, be said against him, and having taken up his position against all gainsayers – knowing that he has sought for objections and difficulties instead of avoiding them, and has shut out no limit which can be thrown upon the subject from any quarter – he has a right to think his judgment better than that of any person, or any multitude, who have not gone through a similar process' (209; my italics). Nor is the insinuation correct that Mill's society is, 'so to speak, a *debating club*' (36, original italics). Just think of Mill's insistence on different *'experiments of living'* (249). Of course, such attention to detail is not to be expected from a self-righteous conservative for whom any discussion of freedom, and any attempt to achieve it, is but 'evil teaching' (35).

The possibilities of Mill's liberalism can be seen from the fact that it provides room for any human desire, and for any human vice. There are no general principles apart from the principle of minimal interference with the life of individuals, or groups of individuals who have decided to pursue a common aim. For example, *there is no attempt to make the sanctity of human life a principle that would be binding for all.* Those among us who can realize themselves only by killing humans and who feel fully alive only when in mortal danger are permitted to form a subsociety of their own where human targets are selected for the hunt, and are hunted down mercilessly, either by a single individual or by specially trained groups (for a vivid account of such forms of life see the film *The Tenth Victim*). So whoever wants to live a dangerous life, whoever wants to taste human blood, will be permitted to do so within the domain of his own subsociety. *But he will not be permitted to implicate others*; for example, he will not be permitted to force others to participate in a 'war of national honor' or what have you. He will not be permitted to cover up whatever guilt he may feel by making a potential murderer out of everyone. It is very strange to see how the *general* idea of the sanctity of human life that frowns upon simple, innocent, and rational murders, such as the murder of a nagging wife by a henpecked husband does not object to the *general* murder of people one has not seen and with whom one has no quarrel. Let us admit that we have different tastes, let those who want to wallow in blood receive the opportunity to do so without giving them the power to make 'heroes' of the rest of society. As far as I am concerned a world in which a louse can live happily is a better world, a more instructive world, a more mature world than a world in which a louse must be wiped out. (For this point of view see the work of Carl Sternheim; for a brief account of Sternheim's philosophy, see Wilhelm Emrich's preface to C. Sternheim *Aus dem Buergerlichen Heldenleben* (Neuwied, 1969), 5–19.) Mill's essay is a first step in the direction of constructing such a world.

It also seems to me that the United States is very close to a cultural laboratory in the sense of Mill where different forms of life are developed and different modes of human existence tested. There are still many cruel and irrelevant restrictions, and excesses of so-called lawfulness threaten the possibilities which the country contains. However, these restrictions, these excesses, these brutalities occur in the brains of human beings; they are not all found in the Constitution. Accordingly, they can be removed by propaganda, enlightenment, special bills, personal effort (Ralph Nader!), and numerous other legal means. Of course, if such enlightenment is regarded as superfluous, if one thinks it irrelevant, if one assumes from the

through that tyranny, that people should be eccentric' (267).[8] Nor should those who 'admit the validity of the arguments for free discussion[s] . . . object to their being "pushed to an extreme" . . unless the reasons are good for an extreme case, they are not good for any case' (210).[9] Thus methodological and humanitarian arguments are intermixed in every part of Mill's essay,[10] and it is on *both* grounds that a pluralistic

very beginning that the existing possibilities for change are either insufficient or condemned to failure, if one is determined to use 'revolutionary' methods (methods, incidentally, which real revolutionaries, such as Lenin, would have regarded as utterly infantile, and which must increase the resistance of the opposition rather than removing it), then, of course, the 'system' will appear much harder than it really is. It will appear harder *because one has hardened it oneself*, and the blame falls back on the bigmouth who calls himself a critic of society. It is depressing to see how a system that has much inherent elasticity is increasingly made less responsive by fascists on the Right and extremists on the Left until democracy disappears without ever having had a chance. My criticism, and my plea for anarchism, is therefore directed *both* against the traditional puritanism in science and society and against the 'new', but actually age-old, antediluvian, primitive puritanism of the 'new' Left which is always based on anger, on frustration, on the urge for revenge, but never on imagination. Restrictions, demands, moral arias, generalized violence everywhere. A plague on both your houses!

[8] For a different argument which is entirely in Mill's spirit, see my 'Problems of Empiricism' *A M*, 185. Today increase of testability can be added to the list of epistemological benefits presented by Mill ('Problems of Empiricism', section 6). This is not a real addition, however, but only a more detailed and more technical presentation of ideas already developed by him.

[9] This quotation has been added mainly for the benefit of Professor Herbert Feigl who keeps making fun of me for adopting extreme positions. Extreme positions are of extreme value. They induce the reader to think along different lines. They break his conformist habits. They are strong instruments for the criticism of what is established and well received. On the other hand, the current infatuation with 'syntheses' and 'dialogues' which are defended in the spirit of tolerance and of understanding can only lead to an end of all tolerance and of all understanding. To defend a 'synthesis' by reference to tolerance means that one is not prepared to tolerate a view that does not show an admixture of one's own pet prejudices. To invite to a 'dialogue' by reference to tolerance means inviting one to state one's views in a less radical and therefore mostly less clear way. An author who can write, in the spirit of 'dialogue', that 'Christianity and Marxism are not contrary to each other' (Guenther Nenning, quoted from the *Newslett. Am. Inst. Marxist Stud.*, 6, no. 1 (1969), 1, bottom) will hardly be prepared to accept the doctrines of a tough-minded Marxist who is interested in progress, not in peace of mind.

[10] In *The Poverty of Liberalism* (Boston, 1968), R. P. Wolff objects to proliferation on the grounds that it does not follow from the happiness principle. This criticism is certainly irrelevant to the thesis of *On Liberty*. The purpose of *On Liberty* is not *to establish a proposition*, be it now by reference to happiness, or in any other way; the purpose is *to set an example*, to present, explain, defend a certain form of life and to show its consequences in special cases (this becomes crystal clear from the relevant pages of the *Autobiography*). True, Mill *also* wrote on the happiness principle; but he was free and inventive enough not to restrict himself to a single philosophy, but to pursue different lines of thought. As a result maximum happiness plays no role in *On Liberty*. What *does* play a role is the free and unrestricted development of an individual. One can understand, however, why the author concentrates on happiness. This gives him the opportunity to display his knowledge (if one can call it that) of some of the tools which analytic 'philosophers' have constructed for the endless discussion of hedonism.

In addition to the complaint just mentioned – for one can hardly call it an argument – Wolff offers what amounts to a series of rhetorical questions. 'It is hard to believe,' he says (17), 'that even the most dedicated liberal will call for the establishing of chairs of astrology in our astronomy departments or insist that medical schools allow a portion of their curriculum to the exposition of chiropractice in order to strengthen our faith in the germ

epistemology is defended, for the natural as well as for the social sciences.[11]

One of the consequences of pluralism and proliferation is that stability of knowledge can no longer be guaranteed. The support that a theory receives from observation may be very convincing; its categories and basic principles may appear well founded; the impact of the evidence may be ex-

theory of disease.' This is hard to believe indeed, for our 'most dedicated liberals' are often moral and intellectual cowards who would not dream of attacking that prize exhibit of the twentieth century, science. Besides, who would think that increasing the number of *university chairs* is going to lead to a more critical point of view? Are university chairs the only things a contemporary 'radical philosopher' (text on front flap of the book) can think on when considering the possibilities of intellectual improvement? Are the limits of a university also the limits of the imagination of our academic radicals? If so, then the attack against Mill collapses at once, for how can a person with such a restricted point of view hope to even *comprehend* the simple non-academic message of Mill's philosophy?

'Does anyone suppose', Wolff continues his inquiry (16), 'that a bright young physicist must keep his belief in quantum mechanics alive by periodically rehearsing the crucial experiments which gave rise to it?' Yes sir, there are lots of people who suppose just that, among them the founders of the quantum theory. There are lots of people who point out that science was often advanced with the help of some *historical* piece of knowledge and who explain the boorishness of much of contemporary physics by the very same lack of perspective which our radical author takes as the basis of his criticism. Of course, 'no material harm' (16) will come from the suppression of history and of alternatives just as brothels do not suffer from the philosophical ignorance of the whores; they flourish, and continue flourishing. But a philosophical courtesan certainly is preferable to a common broad because of the added techniques she can develop; and a science with alternatives is preferable to the orthodoxy of today for exactly the same reasons.

It is interesting to see how conservative so-called 'radicals' become when confronted with the apparently more solid and more difficult parts of the establishment, such as science. Which again shows that they are moral cowards who dare to sing their arias only when there is absolutely no danger of a serious intellectual fight and when they can be absolutely sure of the support of what they think are the 'progressive' elements of society.

[11] Later in the nineteenth century proliferation was defended by *evolutionary arguments*: just as animal species improve by producing variations and weeding out the less competitive variants, science was thought to improve by proliferation and criticism. Conversely, 'well-established' results of science and even the 'laws of thought' were regarded as temporary results of adaptation; they were not given absolute validity. According to L. Boltzmann (*Populäre Schriften* (Leipzig, 1905), 398, 318, 258–9), the latter 'error finds its complete explanation in Darwin's theory. Only what was adequate was also inherited. . . . In this way the laws of thought obtained an impression of infallibility that was strong enough to regard them as supreme judges, even of experience . . . One believed them to be irrefutable and perfect. In the same way our eyes and ears were once assumed to be perfect, too, for they are indeed most remarkble. Today we know that we were mistaken – our senses are not perfect.' Considering the hypothetical status of the laws of thought, we must 'oppose the tendency to apply them indiscriminately, and in all domains' (40). This means, of course, that there are circumstances, not factually circumscribed or determined in any other way, in which we must introduce ideas that contradict them. We must be prepared to introduce ideas inconsistent with the most fundamental assumptions of our science *even before* these assumptions have exhibited any weakness. Even 'the facts' are incapable of restricting proliferation, for 'there is not a single statement that is pure experience' (286, 222). Proliferation is important not only in science but in other domains too: 'We often regard as ridiculous the activity of the conservatives, of those pedantic, constipated, and stiff judges of morality and good taste who anxiously insist on the observance of every and any ancient custom and rule of behavior; but this activity is beneficent and it must be carried out in order to prevent us

tremely forceful. Yet there is always the possibility that new forms of thought will arrange matters in a different way and will lead to a transformation even of the most immediate impressions we receive from the world. Considering this possibility, we may say that the long-lasting success of our categories and the omnipresence of a certain point of view is not a sign of excellence or an indication that the truth or part of the truth has at last been found. It is, rather, the indication of a *failure of reason* to find suitable alternatives which might be used to transcend an accidental intermediate stage of our knowledge. This remark leads to an entirely new attitude toward success and stability.

As far as one can see, the aim of all methodologies is to find principles and facts which, if possible, are not subjected to change. Principles which give the *impression* of stability are, of course, tested. One tries to refute them, at least in some schools. If all attempts at refutation fail, we have a *positive result*, nevertheless: we have succeeded in discovering a new stable feature of the world that surrounds us; we have come a step closer to the truth.

Moreover, the process of refutation itself rests on assumptions which are not further investigated. An instrumentalist will assume that there are stable facts, sensations, everyday situations, classical states of affairs, which do not change, not even as the result of the most revolutionary discovery. A 'realist' may admit changes of the observational matter, but he will insist on the separation between subject and object and he will try to restore it wherever research seems to have found fault with it.[12] Believing in an 'approach to the truth', he will also have to set limits to the development of

from falling back into barbarism. Yet petrification does not set in, for there are also those who are emancipated, relaxed, the hommes sans gêne. Both classes of people fight each other and together they achieved a well-balanced society' (322).

But Boltzmann does not always carry his ideas through to the end. Occasionally he relies on a more simplistic empiricism such as when he says that 'a well-determined fact remains unchanged forever' (343), or when he regards 'my waking sensations [as] the only elements of my thought' (173) so that 'we infer the existence of objects from the impressions made on our senses (19), or when he declares, more than once, that the task of science is 'to adapt our thoughts, ideas, and concepts to the given rather than subjecting the given to the judgment of the laws of thought' (354; cf. with this the assertion, on p. 286, that 'the simplest words such as yellow, sweet, sour, etc. which seem to represent mere sensations already stand for concepts which have been obtained by abstracting from numerous facts of experience'). He also warns us not to 'go too far beyond experience'. This vacillation between a sound scientific philosophy and a bad positivistic conscience is characteristic of almost all so-called 'realists' from Boltzmann up to, and including, Herbert Feigl. Reasons for this phenomenon are found in Lenin's *Materialisam and Empirio-Criticism* (New York, 1927). Hegel's philosophy which tells us why we can and should go as far beyond experience as possible has considerably improved the situation. All that is needed now is a little dialectics and attention to specific historical conditions.

[12] Popper, for example, takes it for granted that the subject cannot enter the domain of science, and he also uses a rather simple form of mechanical materialism in his attack on Bohr. For details see vol. 1, ch. 16, esp. section 9. All these principles are used by him dogmatically, and without the shred of an argument. No Hegelian would ever proceed in such a simple-minded manner.

concepts. For example, he will have to exclude incommensurable concepts from a series of succeeding theories.[13] This is the traditional attitude, up to, *and including*, Popper's 'critical' rationalism.

As opposed to it, a thinker following Mill (or Hegel) regards any prolonged stability, either of ideas and impressions or of background knowledge which one is not willing to give up (realism; separation of subject and object; commensurability of concepts), as an indication of *failure*, pure and simple. Any such stability indicates *that we have failed to transcend an accidental stage of research and that we have failed to rise to a higher stage of consciousness and of understanding*. It is even questionable whether we can still claim to possess knowledge in such a state. As we become familiar with the existing categories and with the alternatives that are being used in the examination of the received view, our thinking loses its spontaneity until we are reduced to the 'bestial and goggle-eyed contemplation of the world around us.'[14] 'The more solid, well defined, and splendid the edifice erected by the understanding, the more restless the urge of life to escape from it into freedom.'[15] Every move away from the status quo, by opening the way to a new and as yet untried system of categories, temporarily returns to the mind the freedom and spontaneity that is its essential property.[16] But complete freedom is never achieved. For any change, however dramatic, always leads to a new system of *fixed* categories. Things, processes, states are still separated from each other. The existence of different elements, of a manifold, is still 'exaggerated into an opposition by the understanding'.[17]

This 'evil manner of reflection,[18] to always work with fixed categories,' (*L*I, 82) is extended by the customary modes of research to the most widely presupposed and unanalysed opposition between a subject and an entirely different world of objects.[19] The following assumptions which are important for a methodological realism have been made in this connection: 'the object . . . is something finished and perfect that does not need the slightest amount of thought in order to achieve reality while thought itself is . . . something deficient that needs . . . material for its completion[20] and must be soft enough to adapt itself to the material in question' (*L*I, 25). 'If thought

[13] Cf. ch. 17 of *AM*.

[14] 'Verhaeltnis des Skeptizismus zur Philosophie' quoted from *Hegel, Studienausgabe* (Frankfurt, (1968) I, 113; see also p. 112.

[15] Hegel, *Differenz des Fichte'schen und Schelling'schen Systems* (Hamburg, 1963), 13.

[16] 'Process becomes converted back to praxis, the patient becomes an agent' R. D. Laing, *The Politics of Experience*, (Baltimore, 1964) 35. There is a good deal of similarity between Hegel's attempt to set concepts in motion and the attempts of some contemporary psychiatrists to return to the individual the control of some of the defence and projection mechanisms he has himself invented.

[17] Hegel, *Logik* II, 61 (hereafter referred to as *L*II).

[18] 'Reflective reason . . . is nothing but the understanding which uses abstraction, separates, and insists that the separation be maintained and taken seriously.' Hegel, *Logik* (Hamburg, 1934), I, 26 (hereafter referred to as *L*I).

[19] Cf. Hegel, *Differenz*, 14.

[20] Cf. the Carnap quotation in vol. 1, ch. 4.

and appearance do not completely correspond to each other, one has, to start with, a choice: the one or the other may be at fault. [Scientific empiricism] blames thought for not adequately mirroring experience . . .'[21] 'These are the ideas which form the core of our customary views concerning the relation between subject and object' (*L*I, 25), and they are responsible for whatever immobility remains in science, even at a time of crisis.

How can this immobility be overcome? How can we obtain insight into the most fundamental assumptions, not only of science and commonsense, but of our existence as thinking beings as well? Insight cannot be obtained as long as the assumptions form an unreflected and unchanging part of our life. But, if they are allowed to change, can we then finish the task of criticizing as identically the same persons who started it? Problems like these are raised not only by the abstract question of criticism, but also by more recent discoveries in anthropology, history of science, and methodology. I have briefly dealt with them when discussing incommensurable theories. For the moment, I would like to indicate, very briefly, how certain ideas of *Hegel* can be used to get a tentative first answer, and thus to make a first step in our attempt to reform the sciences.

Science, commonsense, and even the refined commonsense of critical rationalism use certain fixed categories ('subject'; 'object'; 'reality'), in addition to the many changing views they contain. They are therefore not fully rational. Full rationality can be obtained by extending criticism to the stable parts also. This presupposes the invention of alternative categories and their application to the whole rich material at our disposal. The categories, and all other stable elements of our knowledge, must be set in motion. 'Our task is to make fluid the petrified material which we find, and to relight [*wieder entzuenden*] the concepts contained in this dead stuff' (*L*II, 211). We must 'dissolve the opposition of a frozen subjectivity and objectivity and comprehend the origin of the intellectual and real world as a becoming, we must understand their being as a product, as a form of producing.'[22] Such dissolving is carried out by reason, which is 'the force of the negative absolute, that is, an absolute negation,'[23] that 'annihilates'[24] science and commonsense, and the state of consciousness associated with both. This annihilation is not a conscious act of a scientist who has *decided* to eliminate some basic distinctions in his field. For although he may con-

[21] *Encyclopädie der Philosophischen Wissenschaften*, ed. G. Lasson (Leipzig, 1920), 72–3. In the original the reference is to Kant, not to scientific empiricism.

[22] *Differenz*, 14. Cf. Lenin's comments on a similar passage in his notes on Hegel's *Logik*, quoted in V. I. Lenin, *Aus Dem Philosophischen Nachlass* (Berlin, 1949), 136ff, esp. 142.

[23] Cf. also 'Skepticismus' Hegel, *Studienausgabe*, 117: 'that scepticism is intrinsically connected with every true philosophy.' Also p. 118: 'Where can we find a more perfect and independent document and system of true scepticism than in Plato's . . . Parmenides? Which embraces and destroys the whole domain of a knowledge achieved by the concepts of our understanding.'

[24] *Differenz*, 25.

sciously try to overcome the limitations of a particular stage of knowledge, he may not succeed for want of objective conditions (in his brain, in his social surroundings, in the physical world[25]) favouring his wish.[26] Hegel's general theory of development, his cosmology, as one might call it, gives an account of such conditions.

According to this cosmology, every *object*, every determinate being, is related to everything else: 'a well determined being, a finite entity is one that is related to others; it is a content that stands in the relation of necessity to another content and, in the last resort, to the world. Considering this mutual connectedness of the whole, metaphysics could assert . . . the tautology that the removal of a single grain of dust must cause the collapse of the whole universe' (*L*I, 71). The relation is not external. Every process, object, state, etc. actually *contains* part of the nature of every other process, object, state, etc.[27] *Conceptually* this means that the complete description of an object is self-contradictory. It contains elements which *say what the object is*; these are the elements used in the customary accounts provided by science and by commonsense which consider part of its properties only and ascribe the rest to the outside. And the description also contains other elements which say *what the object is not*. These are the elements which science and commonsense put outside the object, attributing them to things which are supposed to be completely separated but which are actually contained in the object under consideration. The result is that 'all things are beset by an internal contradiction' (*L*II, 53). This contradiction cannot

[25] 'It is my aim to read Hegel in a materialistic fashion . . .' Lenin, *Nachlass*, 20. The same is true of Professor D. Bohm.

[26] Cf. the note on the limit and the ought, *L*I, 121–2: 'Even a stone, being something, is differentiated into its being for itself and its Being and so it, too, transcends its limit . . . If it is a basis for acidification, then it can be oxidized, neutralized, and so on. In the process of oxidation, neutralization, etc. its limit, i.e., only to be a basis, is removed . . . and it contains the *ought* to such an extent that only force can prevent it from ceasing to be a basis . . .'

[27] 'Everything that exists is linked in this way to everything else: to the *total process* of the universe. This linkage is either direct, by means of a single quantum, or else indirect, through a series of such linkages.' This is how Bohm describes (*Scientific Change*, ed. A. Crombie (London, 1963), 478) the situation created by the quantum theory. The similarity to Hegel is no accident. Bohm has studied Hegel in detail, and he has taken the *Logik* especially as the point of departure for some of his scientific views: '. . . may we not try to understand the world as a total process, in which all parts (for example, the system under observation, observing apparatus, man, etc.) are aspects or sides whose relationships are determined by the way in which they are generated in the process? Of course, in physics, man can, in an adequate approximation, probably be left out of the totality, because he obtains his information from a piece of apparatus on the large-scale level, which is influenced in a negligible way by his looking at it. But at a quantum mechanical level of accuracy, the apparatus and the system under observation must be recognized to be linked indivisibly. Should not the theory be formulated so as to say that this is so . . . ? In a total process of the kind that I am talking about, an observation is regarded as a particular kind of movement, in which some aspects of the process are, as it were, 'projected' into certain large-scale results . . . This process of projection is . . . an integral part of the total process that is being projected' (482).

be eliminated by using different words, for example, by using the terminology of a *process* and its *modifications*. For the process will again have to be separated, at least in thought, from something other than itself; otherwise it is *pure* being which is in no way different from pure *nothingness*.[28] It will contain part of what it is separated from, and this part will have to be described by ideas inconsistent with the ideas used for describing the original process, which therefore is bound to contain contradictions also.[29] Hegel has a marvellous talent for making visible the contradictions which arise when we examine a concept in detail, wishing to give a complete account of the state of affairs it describes. 'Concepts which usually appear stable, unmoved, dead are analyzed by him and it becomes evident that they move.'[30]

Now we come to a second principle of Hegel's cosmology. The motion of concepts is not merely a motion of the *intellect*, which, starting the analysis with certain determinations, moves away from them and posits their negation. It is an *objective* development as well, and it is caused by the fact that every finite (well-determined, limited) object, process, state, etc., has the tendency to emphasize the elements of the other objects present in it, and to become what it is not. The object, 'being restless within its own limit,' (*L*I, 115) 'strives *not* to be what it is'.[31] 'Calling things *finite*, we want to say that they are not merely limited . . . but rather that the negative is essential to their nature and to their being . . . Finite things *are*, but the truth of their being is their *end*.[32] What is finite does not merely change . . . it passes away; nor is this passing away merely possible, so that the finite thing could continue to be, without passing away; quite the contrary, the being of a finite thing consists in its having in itself the seeds of passing away . . . the hour of its birth is the hour of its death' (*L*I, 117). 'What is finite, therefore, can be set in motion' (*L*I, 117).

Moving beyond the limit, the object ceases to be what it is and becomes what it is not; it is *negated*. A third principle of Hegel's cosmology is that the result of the negation is 'not a mere nothing; it is a *special* content, for . . . it is the negation of a determined and well defined thing' (*L*I, 36). Conceptually speaking, we arrive at a 'new concept which is higher, richer than the concept that preceded it, for it has been enriched by its negation or opposition, contains it *as well as* its negation, being the unity of the original

[28] *L*I, 67. Cf. also the physical model for this identity in *L*I, 78–9, according to which neither 'pure light' nor 'pure darkness' gives rise to (the perception of) objects which are recognized and 'distinguished only in the determined light . . . which is turbid light'.

[29] Bohm will therefore not be able to keep contradiction out of his ideas as he occasionally seems to believe (e.g. in *Scientific Change*, 482, second paragraph). He agrees in other places but tries to circumvent particular contradictions by moving to a different level of reality. Cf. his *Causality and Chance in Modern Physics* (New York, 1961).

[30] Lenin, *Nachlass*, 27.

[31] Hegel, in *Jenenser Logik, Metaphysik und Naturphilosophie*, ed. G. Lasson (Hamburg Felix Meiner, 1967), 31.

[32] In German the statement is more impressive: 'Die Wahrheit [des] Seins der *endlichen* Dinge ist ihr *Ende*.'

concept and of its opposition' (*L*I, 36). This is an excellent description, for example, of the transition from the Newtonian conception of space to that of Einstein, *provided* we continue using the *unchanged* Newtonian concept. 'It is clear that no presentation can be regarded as scientific that does not follow the path and simple rhythm of this method, for this is the path pursued by the things themselves.'[33]

Considering that the motion beyond the limit is not arbitrary, but is directed 'towards its [i.e. the object's] end' (*L*I, 117) it follows that not *all* the aspects of other things which are present in a certain object are realized in the next stage. Negation, accordingly, 'does not mean simply saying No, or declaring a thing to be non-existent, or destroying it in any way one may choose . . . Each kind of thing . . . has its own peculiar manner of becoming negated, and in such a way that a development results from it, and the same holds good for each type of ideas and conceptions . . . *This must be learned like everything else.*'[34] What has to be learned, too, is that the 'negation of the negation' does not lead further away from the original starting point but that it returns to it (*L*I, 107). This is an 'extremely universal and just on that account extremely far-reaching and important law of development in nature, history and thought; a law which . . . asserts itself in the plant and animal world, in geology, in mathematics,[35] in history, in philosophy'.[36] Thus for example 'a grain of barley falling under suitable conditions on suitable soil disappears, is negated, and in its place there arises out of it the plant, the negation of the grain . . . This plant grows, blossoms, bears fruit and finally produces other grains of barley, and as soon as these ripen, the stalk dies, is in turn negated. As a result of this negation of the negation, we again have the grain of barley we started with, not singly, but rather in ten,

[33] *L*I, 36; cf. also *L*II, 54, 58ff.

[34] F. Engels, *Anti-Duehring* (Chicago, 1935), 144–5; my italics. I am quoting Engels, Lenin, Mao, and similar thinkers rather than the usual bunch of Hegelian or anti-Hegelian scholars as they have still kept the freshness of mind that is necessary to interpret *and to apply concretely* the Hegelian philosophy. The same is true of such physicists as Bohm, Vigier, and even Bohr who may occasionally be regarded as an unconscious Hegelian. Cf. the remarks on subject and object below.

[35] Mathematics was for a long time regarded as lying outside the domain of dialectics. The examples used by Hegel and Engels and especially the example of the differential calculus, so it was thought, only showed the immaturity of the mathematics of the time and the limitations of even the greatest philosophers. One should not have been quite so generous, however. What Hegel says of mathematics applies to *informal* mathematics and, insofar as informal mathematics is the source of the rest, to all of mathematics. That a dialectical study of mathematics can lead to splendid discoveries, even today, is shown by Lakatos' *Proofs and Refutations* (Cambridge, 1978). One must praise Lakatos for having made such excellent use of his Hegelian upbringing. On the other hand one must perhaps also criticize him for not revealing his source of inspiration in a more straightforward manner but giving the impression that he is indebted to a much less comprehensive and much more mechanical school of thought. Or has his temporary membership in this school made him lose his sense of perspective; so that he prefers being mistaken for a Wittgensteinian to being classified with the dialectical tradition to which he belongs?

[36] Engels, *Anti-Duehring*, 143–4.

twenty or thirtyfold number . . . and perhaps even qualitatively improved . . .'[37] 'It is evident that I say nothing whatever about the particular process of development which, for example, the grain of barley undergoes from its germination up to the dying off of the fruit-bearing plant, when I state that it is the negation of the negation . . . I rather comprise these processes altogether under this one *law of motion* and just for that reason disregard the peculiarities of each special process. *Dialectics*, however, is nothing else than the science of the general laws of motion and development in nature, human society and thought.'[38]

In the foregoing account, concepts and real things have been treated as separate. Similarities and correspondences were noted: each thing *contains* elements of everything else, it *develops* by turning into these alien elements, it *changes*, and it finally tries to *return* to itself. The *notion* of each thing, accordingly, contains contradictory elements. It is negated, and it moves in a way corresponding to the movement of the thing. This presentation has one serious disadvantage: 'Thought is here described as a mere subjective and formal activity while the world of objects, being situated *vis-à-vis* thought, is regarded as something fixed and as having independent existence. This dualism . . . is not a true account of things and it is pretty thoughtless to simply take over the said properties of subjectivity and objectivity without asking for their origin . . . Taking a more realistic view we may say that the subject is only a stage in the development of being and essence.'[39] The concept, too, is then part of the general development of nature, in a materialistic interpretation of Hegel. 'Life', for example, 'or organic nature is that phase of nature when the concept appears on the stage; it enters the stage as a blind concept that does not comprehend itself, i.e., does not think' (*L*II, 224). Being part of the *natural behaviour*, first of an organism, then of a thinking being, it not only mirrors a nature that 'lies entirely outside of it', (*L*II, 227) it is not merely 'something subjective and accidental', (LII, 408) it is not 'merely a concept' (*L*II, 225); it participates in the general nature of all things, i.e. it contains an element of everything else, it has the tendency to be the end result of the development of a specific thing, so that, finally, the concept and this thing become one (*L*II, 408).

[37] *Ibid.*, 138–9.

[38] *Ibid.*, 144; my italics. Epistemologically these laws belong to the Aristotelian rather than to the Newtonian tradition.

[39] *Encyclopädie der Philosophischen Wissenschaften, ergaenzt durch Vortraege und Kollegienhefte*, ed. L. Henning *et al.* (Berlin, 1840), 395–6; cf. also Lenin, *Nachlass*, 102. Or, to use Bohm's terminology 'as long as, by our customary habits of thinking, we try to say that in an experiment, some part of the world is observed [and described], with the aid of some other part, we introduce an element of confusion into our thought process. Indeed, even the very word "observation" is misleading, as it generally implies a separation between the observing apparatus and the object under observation, of a kind that does not actually exist.' *Scientific Change*, 482–3. The reader should go on and consider the beautiful example of the observation of a mirror image.

'That real things do not agree with the idea [read: 'with the total knowledge of man'[40]] constitutes their *finitude*, their *untruth* because of which they are *objects*, each determined in its special sphere by the laws of mechanics, chemistry, or by some external purpose' (*L*II, 410). In this stage 'there can be nothing more detrimental and more unworthy of a philosopher than to point, in an entirely vulgar fashion, to some experience that contradicts the idea . . . When something does not correspond to its concept, it must be led up to it' (*L*II, 408–9; counterinduction!) until 'concept and thing have become one'.[41]

To sum up: knowledge is part of nature and is subjected to its general laws. The laws of dialectics apply to the motion of objects and concepts, as well as to the motion of higher units comprising objects and concepts. According to these general laws, every object participates in every other object and tries to change into its negation. This process cannot be understood by attending to those elements in our subjectivity which are still in relative isolation and whose internal contradictions are not yet revealed. (Most of the customary concepts of science, mathematics, and especially the rigid categories used by our modern axiomaniacs are of this kind as has already been observed by the Hegelian Lakatos.[42]) To understand the process of negation we must attend to those other elements which are fluid, about to turn into their opposites, and which may, therefore, bring about knowledge and truth, 'the identity of things and concepts'.[43] The identity itself cannot be achieved mechanically, i.e. by arresting some aspect of reality and fiddling about with the remaining aspects, or theories, until agreement is achieved (the aspects one wants to arrest, being in motion, will soon be replaced by dogmatic opinions of them, rigid perceptions included). We must rather proceed dialectically, i.e. by an *interaction* of concept and fact (observation, experiment, basic statement, etc.) that effects *both* elements. The lesson for methodology is, however, this: Do not work with stable concepts. Do not eliminate counterinduction. Do not be seduced into thinking that you have at last found the correct description of 'the facts' when all that has happened is that some new categories have been adapted to some older forms of thought, which are so familiar that we take their outlines to be the outlines of the world itself.

[40] Lenin, *Nachlass*, 114.

[41] *L*II, 228. 'Knowledge is the eternal infinite approach of thought and object. The mirroring of nature in human thought is not "dead," it is not "abstract," it is not without motion, not without its contradictions but is to be conceived as an eternally moving process that gives rise to contradictions and removes them.' Lenin, *Nachlass*, 115.

[42] Cf. his *Proofs and Refutations*.

[43] *L*II, 228. The whole introduction to the *Subjective Logik*, i.e. *L*II, 213–23, can be used for a criticism of what has become known as Tarski's theory of truth. If I remember correctly, this criticism is similar to a criticism voiced by the late Professor Austin in his lectures in Berkeley in 1959, which shows that even an Oxford philosopher occasionally stumbles upon The Truth.

5

Philosophy of science versus scientific practice

Observations on Mach, his followers and his opponents

Modern philosophy of science arose from the Vienna Circle and its attempt to reconstruct the rational components of science. It is interesting to compare its approach with that of earlier philosophers, for example Ernst Mach.

Ernst Mach was a scientist. He was an expert in physics, psychology, physiology, the history of science and the general history of ideas. Ernst Mach was also an educated man. He was familiar with the arts and the literature of his time and he was interested in politics. Even when already paralysed he had himself wheeled into a session of parliament to cast his vote in connection with workers' legislation.

Ernst Mach was not satisfied with the science of his time. As he saw it science had become partially petrified. It used entities such as space and time and objective existence but without examining them. Moreover philosophers had tried to show, and scientists had started believing, that these entities could not be examined by science because they were 'presupposed' by it. This Mach was not prepared to accept. For him every part of science, 'presuppositions' included, was a possible topic of research and subject to correction.

On the other hand it was clear that the correction could not always be carried out by means of the customary procedures, which contained some ideas in a way that protected them from difficulties. It was therefore necessary to introduce a new type of research based on a new cosmology. Mach gave a rough outline of what it would assume and how it would proceed.

According to Mach science deals with *elements and their relations*. The nature of the elements is not given but must be discovered. Known things such as sensations, physical objects, systems of physical objects in space are combinations of elements. The combinations may reproduce the old distinctions, but they may also lead to arrangements of an entirely different kind; for example, they may lead to an interpenetration of 'subject' and 'object' in the old sense. Mach was convinced that the old distinctions were inadequate and had to be given up.

Mach's conception of science has two features that distinguish it from that of today's philosophers of science.

First, Mach was critical towards science *as a whole*.[1] Modern philosophers occasionally make a big show of their independence and their expert knowledge by criticizing particular scientific theories and suggesting minor changes. But they would never dare to criticize science as a whole. They are its most obedient servants. Secondly, Mach criticized scientific ideas not by comparing them with external standards (criteria of meaning or demarcation) but by showing how *scientific research itself* suggested a change. For example, methodological principles were examined not by consulting an abstract and independent theory of rationality but by showing how they aided or hindered scientists in the solution of concrete problems. (Later on Einstein and Niels Bohr developed this procedure into a fine art.)

A third interesting feature of Mach's 'philosophy'[2] was its disregard for distinctions between areas of research. Any method, any type of knowledge could enter the discussion of a particular problem. In building up his new science Mach appealed to mythology, physiology, psychology, history of ideas, history of science as well as to the physical sciences. The *magic world view* which he received from Tylor and Frazer dissolves the distinction between subject and object without ending up in chaos. Mach did not accept this world view, but he used it to show that the nineteenth-century idea of objective existence was not a *necessary* ingredient of thought and perception. His detailed studies in the physiology of the senses showed him that it was not *adequate* either. Sensations are complex entities containing 'objective' ingredients, 'objects' are constituted by processes (e.g. Mach bands) which belong to the 'subject', the boundary between subject and object changes from one case to the next: for us it lies at our fingertips, for the blind man using a stick it lies at the end of the stick. The *history of science* and *physics* showed that 'objective' theories such as Newton's theory of space and time and atomism were in trouble precisely because of their objectivist features. On the other hand, there existed theories of a different kind such as the phenomenological theory of heat which were successful though not based on material substances. Having started with a classification of sensations (cf. Mach's *Theory of Heat* (German edition, Leipzig, 1896)) such theories suggested to Mach that *at least at this stage of research* elements could be identified with sensations. For the time being, Mach's new science could therefore be developed from two assumptions:

(1) the world consists of elements and their relations. The nature of the elements and of the relations as well as the way in which things are built up from them is to be determined by research using the concepts that seem most economical at a certain stage of science, and

[1] See his debate with Planck, reprinted in *Physical Reality*, ed. S. Toulmin (New York, 1968).

[2] I put the word in quotation marks because Mach always refused to be regarded as the proponent of a new 'philosophy', which agrees with the above account of this research practice.

(2) elements are sensations.

This is how Mach combined the information provided by different fields to give shape to his own idea of research.[3]

Mach's idea of research was more comprehensive than that of his contemporaries and certainly of all his philosophical successors. Before Mach's time it had been taken for granted that not all parts of science could be examined by scientific means. Space, time, observer independence were thought to be beyond the reach of (scientific) argument. Now there were means of criticizing not only these ideas but the very standards of research: no standards can guide research without being subjected to the control of research.

It is interesting to see how later 'scientific' philosophers changed this rich and fruitful point of view. Mach's attempt to make research more comprehensive so that it could deal with 'scientific' as well as 'philosophical' questions remained unnoticed both by his followers and by his opponents. What they did notice were his assumptions and hypotheses and those they turned into 'principles' of precisely the kind Mach had rejected. The theory of elements became a 'presupposition', the identification of elements and sensations a definition, and relations between concepts were imposed in accordance with some rather simpleminded rules; they were no longer determined by research. The construction of conceptual systems having such rules and such principles as their boundary conditions now became *the* task of a new and rather aggressive discipline, the philosophy of science. With this the old dichotomy between philosophical speculation and scientific research which Mach had tried to absorb into science reappeared, but it was a very impoverished and illiterate philosophy that took the place of its glorious ancestors. Being contemptuous of earlier ideas the new philosophers lacked perspective and soon repeated all the traditional mistakes.[4]

[3] Note that the second assumption above is a *hypothesis* and not a 'presupposition' of research. It is comparable with the assumption of 'objectivist' scientists that the ultimate building blocks of matter are elastic spheres, like billiard balls. It gets research started, it is not an unchanging standard of its adequacy. Having criticized the standards and 'presuppositions' of the science of his time Mach was not going to replace them by some other dogmatism (this becomes very clear from his notebooks).

[4] A predicament they shared with the enlightenment, except that the writers of that period *invented* their philosophy while the members of the Vienna Circle just *copied* the distorted ideas of their great predecessors. Also philosophers of the enlightenment dealt with ethics, aesthetics, theology, they founded a new anthropology and considerably widened the horizon of their contemporaries. Nothing even slightly comparable is offered by the new 'scientific' philosophy that came out of the Vienna Circle (and Popperianism) and which is mostly concerned with the physical sciences and some distorted images of man, as we have seen. Any extension beyond these boundaries is second-hand imitation of earlier views and shares the superficiality of such imitations. Characteristic for the enterprise is a schoolmasterly tone which occurs wherever imagination has ceased to work and has been replaced by routine responses. A superficial comparison between Popper and, say, Lessing shows the difference

There then arose again two ways of dealing with general problems such as problems of space, time, reality and related problems viz. *the way of the scientists* and *the way of the philosophers*.

A *scientist* starts with a bulk of material consisting of diverse and conflicting ingredients. There are theories formulated in accordance with the highest standards of rigour and precision side by side with unfounded and sloppy approximations,[5] there are 'solid' facts, local laws based on some of these facts, there are heuristic principles, tentative formulations of new points of view which partly agree, partly conflict with the accepted facts, there are vague philosophical ideas, standards of rationality and procedures that conflict with these. Being unable to make such material conform to simple views of order and consistency the scientist usually develops a *practical logic* that permits him to get results amidst chaos and incoherence. Most of the rules and standards of this practical logic are conceived *ad hoc*; they serve to remove a particular difficulty and it is not possible to turn them into an organon of research. 'The external conditions' writes Einstein[6] 'which are set for [the scientist] . . . do not permit him to let himself to be too much restricted, in the construction of his conceptual world, by the adherence to an epistemological system. He therefore must appear to the systematic epistemologist as a type of unscrupulous opportunist . . .' And Niels Bohr 'would never try to outline any finished picture, but would patiently go through all the phases of a problem, starting from some apparent paradox, and gradually leading to its elucidation. In fact he would never regard achieved results in any other light than as starting points for further exploration. In speculating about the prospects of some line of investigation, he would dismiss the usual considerations of simplicity, elegance, or even consistency with the remark that such qualities can only be properly judged after the event . . .[7] It is of course possible to describe particular cases but the only lesson we can draw from the descriptions is cautionary: never expect a clever trick, or a 'principle', that was helpful on one occasion to succeed on another. One outstanding feature of scientific research, especially of the kind envisaged by Mach, is its disregard for established boundaries. Galileo argued as if the distinction between astronomy and physics which was a basic presupposition of the knowledge of his time did not exist; Boltzmann used considerations from mechanics, the phenomenological theory of heat and optics to determine

between true enlightenment and the slavish imitation of its outer form. (Kant who wanted to become famous and who knew that schoolmasters are more readily accepted than independent minds changed his style in midlife. And he was right: the three *Critiques* became a great success.)

[5] Cf. the account of *ad hoc* approximations in *AM*, 63.

[6] *Albert Einstein: Philosopher-Scientist*, ed. P. A. Schilpp (New York, 1951), 683ff.

[7] L. Rosenfeld in *Niels Bohr, his Life and Work as seen by his Friends and Colleagues*, ed. S. Rosenthal (New York, 1967), 117.

the scope of the kinetic theory; Einstein combined specific approximations with a global and very 'transcendental' survey of physical world views; Heisenberg got some of his basic ideas from the *Timæus* and, later on, from Anaximander. Metaphysical principles are used to advance research; logical laws and methodological standards are suspended without much ado as constituting undue restrictions; adventurous and 'irrational' conceptions abound. The successful researcher frequently is a literate man, he knows many tricks, ideas, ways of speaking, he is familiar with details of history and abstractions of cosmology, he can combine fragments of widely differing points of view and quickly switch from one framework to another. He is not tied to any particular language for he may speak the language of fact and the language of fairytale side by side and mix them in the most unexpected ways. And, mind you, this applies both to the 'context of discovery' *and* to the 'context of justification', for examining ideas is just as complex an activity as introducing them.

The strife about the kinetic theory of matter towards the end of the last century and the rise of the quantum theory are excellent examples of the features I have just described. In the case of the quantum theory we have classical celestial mechanics, classical electrodynamics and the classical theory of heat. Sommerfeld and Epstein strained the first and the second to the limit by supplementing them with a 'fourth Keplerian law' viz. the quantum conditions. Their success suggested that quantum mechanics might be developed out of classical theory without too much change. On the other hand the original considerations of Planck, generalized by Poincaré, seemed to indicate that fundamental ideas such as the idea of a trajectory were inherently problematic. Einstein, recognizing their problematic character, worked almost entirely with approximations and inferences from them and his results (photoelectric effect; statistical studies) had only a limited application: they could not explain the laws of interference. They even seemed to clash with experiments and received scant attention until Millikan showed the correctness of some of the predictions made. Working with approximations then became *the* method of the Copenhagen school. The method was disliked and not well understood by physicists of Sommerfeld's persuasion but explained the limited applicability of even the most subtle mathematical instruments. And just as a great and troublesome river deposits many strange objects on its shores, in the very same fashion the great and troublesome river of pre-1930 quantum mechanics produced numerous precise but little understood results, both in the form of 'facts' and in the form of 'principles' (Ehrenfest's principle of adiabatic change being one of them).

The *way of the philosopher* is very different; there could not be a greater contrast. There are some general ideas and standards which are spelled out in detail and there are the principles of the logic chosen. There is hardly anything else – a consequence of the 'revolution in philosophy' initiated by the Vienna Circle. The logic used was of course discussed and it changed, for logic is a science like every other science – but only its most pedestrian parts entered the philosophical debate. Thus we have not only a separation between science and philosophy, but a further separation between a scientific ('mathematical') logic and a logic for philosophers. It is as if scientists used not the most advanced mathematics of their time but some backward idiom, and tried to formulate their problems in its terms. Research of the philosophical type then consists in proposing ideas that fit the boundary conditions, i.e. the standards and the simple logic chosen.

Such ideas clearly are both too wide and too narrow. They are too wide because the contemporary knowledge of facts is not taken into account (a purely philosophical theory of walking is bound to be too wide because it does not consider the restrictions imposed by physiology and landscape). And they are too narrow because the restricting standards and rules are also unaffected by that knowledge (a purely philosophical theory of walking is too narrow because it imposes restrictions not paralleled by the vast possibilities of human motion). It is this last feature that makes a philosophical criticism so dreary and repetitive. While a good scientist objects to 'making a good joke twice'[8] a philosopher insists on standard arguments against standard violations of standard standards. Exclamations such as 'inconsistent!', '*ad hoc*!', 'irrational!', 'degenerating!', 'cognitively meaningless!' recur with tiring regularity. Illiteracy, however, not only does not matter, it is a sign of professional excellence. It is *required*, not just tolerated. All the distinctions of the discipline (context of discovery/context of justification; logical/psychological; internal/external; and so on) have but one aim: to turn incompetence (ignorance of relevant material and lack of imagination) into expertise (happy assurance that the things not known and unimaginable are not relevant and that it would be professionally incompetent to use them).

The much admired addition of modern formal logic to philosophy has encouraged the illiteracy by providing it with an organon. More than anything else it enabled the barren fathers of positivism to deny their shortcomings and to assert, not without considerable pride, that they were not concerned with the advancement of knowledge but with its 'clarification' or its 'rationality'. Even critics did not try to re-establish contact with the practice of science,[9] they merely tried to free the suggested 'reconstructions' from internal difficulties.[10] The distance between scientific practice

[8] Einstein's reply to the question why he did not stick to the philosophical ideas that led him to special relativity. See Philipp Frank, *Einstein, His Life and Times* (London, 1948), 261.

[9] Lakatos, of course, tried to find a connection, but he came later and succeeded in making only *verbal* contact; cf. *A M*, 196ff.

[10] Thus Popper's theory of falsification concerns an improvement of *confirmation logic*, not of science. The same is true of his theory of verisimiltude.

and the philosophy of science remained as large as ever. But this deficiency, this astounding unreality of the enterprise had already been turned into an asset: differences between reconstructions and actual science were regarded as faults of *science*, not as faults of the reconstructions. Of course, nobody was bold enough to play this game with physics (though there were some who derived much mileage from conflicts *inside* physics such as the conflict between Bohr and Einstein); but if the trouble came from a less adored science then the verdict was clear: off with her head! While Mach's criticism was part of a *reform* of science that *combined criticism with new results*, the criticism of the positivists and of their anxious foes the critical rationalists proceeded from some frozen ingredients of the Machian philosophy (or modifications of it) that could no longer be reached by the process of research. Mach's criticism was dialectical and fruitful, the criticism of the philosophers was dogmatic and without fruit. It mutilated science instead of making it grow. This started the trend whose late children we have now before us.

It is interesting to compare the two procedures in a concrete case.[11]

Mach's idea of a science all of whose standards and principles are under its own control was realized, in different ways, by Einstein and by Bohr. Interestingly enough both scientists (and some of their followers such as Max Born) regarded themselves as dilettantes; they defined and approached their problems independently of existing standards. They had no compunction about mixing science and philosophy and so advanced the course of both. Einstein's philosophical bent becomes clear from the way in which he arranges his material, Bohr's philosophy is an essential element of the older quantum theory.[12] It is true that Mach was severely critical of some later consequences of Einstein's research, but one should examine his reasons before concluding that Einstein was outside Mach's research programme. Nobody has so far paid attention to Mach's remark, contained in his criticism, that his investigations in the physiology of the senses had led him to results different from those ascribed to relativity. This establishes a connection with Mach's earlier analysis of space and time and indicates that he did not object to the new *theory* but to its *reification* by Planck and von Laue. For here relativity was used to support the very same naive and inarticulate notion of reality that Mach had objected to and had started to examine. The examination was continued by the quantum theory which gave new content to the notion of an element, revealed new and complex relations between elements, and so modified our idea of reality. All this

[11] In the Vienna Circle only Neurath had a clear conception of the properties of scientific research (as opposed to philosophical analysis). The difference between the two modes is well explained in A. J. Ayer's criticism of Neurath in his *Foundations of Empirical Knowledge* (London, 1964).

[12] For details cf. vol. 1, ch. 16.

occurred in the twenties and thirties. What did the philosophers have to offer during this time and after?

They had very little to offer in the case of relativity. They watched the development from the wings, applauded it, and 'clarified' it, i.e. described it as an instance of problem solving in their sense. The 'clarification' created some interesting philosophical myths. For example, it created the myth that Einstein advanced by eliminating metaphysics, or by eliminating *ad hoc* hypotheses, or because he was an operationalist, or because he took refutations seriously. Zahar's remark that special relativity was no advance at all is the latest and most amusing myth of this kind.

The situation was different in the case of the quantum theory and its conception of 'reality'. While the quantum race was on the Vienna Circle changed from sense-data languages to physicalistic languages. The change was as arbitrary as the choice of sense data had been in the first place. Sense data were removed because one returned to an interpretation of science that the idea of sense data had been supposed to test. And one returned to such an interpretation not because a test was performed and failed – one was not even aware of the test function of sense data in Mach's philosophy – one simply remembered some principles of the science he wanted to improve and used them as arguments against carrying out the improvement.[13] This rather unreflected turnabout occurred at precisely the time when the idea of objective existence was examined by physicists and replaced by a more complex account of reality. The event had no effect whatsoever on the debates between physicalists and the defenders of sense data, and one can see why. It was regarded not as introducing different, more complex and more realistic arguments about the issue but simply as a technical but philosophically inferior version of an examination of the second, the philosophical kind. This becomes very clear from Popper's account of the matter. Writing more than twenty years later he complains 'Without any debate over the philosophical issue, without producing any new argument, the instrumentalist view has become an accepted dogma'.[14] For him the detailed physical arguments, the many attempts to escape 'instrumentalism' as he calls the final position of the Copenhagen school simply do not exist. The tendency to 'translate' cosmological assumptions into the 'formal mode of speech' and so to conceal their factual content aided this blindness and the resulting rigidity of their philosophers' approach. Thus Popper, in the essay already quoted[15] removes 'essentialism' and introduces 'realism' by referring to the 'fact' (as he calls it) 'that the world of each of our theories

[13] The arbitrary nature of the change is clearly realized by Carnap who in his *Logical Syntax of Language* (German edition, Vienna, 1934) and earlier papers makes the choice of sense data versus physicalistic languages a matter of *convenience* ('principle of tolerance').

[14] *Conjectures and Refutations* (London, 1963), 99f.

[15] *Ibid.*, 115.

may be explained by further worlds . . . described by further theories'. This, of course, is his model of science in which rejection of *ad hoc* hypotheses plays an essential role. The model breaks down in a finite world, but the breakdown will never become visible to a philosopher who hides factual assumptions behind 'logical' principles and 'methodological' standards. This is how complex problems needing unusual ideas and unusual minds were turned into trite puzzles which were then explained at great length and solved with a great show of intellectual effort.[16] And this is how the reality conception of classical physics was able to stage a comeback in philosophy after it had been defeated by scientific research.

The writers of the Vienna Circle and the early critical rationalists who distorted science and ruined philosophy in the manner just described belonged to a generation still vaguely familiar with physics. Besides, they started a new trend; they did not merely take it over from more inventive predecessors. They *invented* the errors they spread, they had to *fight* to get them accepted and so they had to possess a modicum of *intelligence*. They also suspected that science was more complex than the models they proposed and so they worked hard to make them plausible. They were pioneers, even if only pioneers of simplemindedness. The situation is very different with the new breed of philosophers of science that now populate our universities. They received their philosophy ready made, they did not invent it. Nor do they have much time or inclination to examine its foundations. Instead of bold thinkers who are prepared to defend implausible ideas against a majority of opponents we have now anxious conformists who try to conceal their fear (of failure, of unemployment) behind a stern defence of the status quo. This defence has entered its epicyclic stages: attention is directed to details and considerable work is done to cover up minor faults and deficiencies. But the basic illiteracy remains and it is reinforced, for hardly anyone of the new breed possesses the detailed knowledge of scientific procedure that occasionally made their ancestors a little hesitant in their pronouncements. For them 'science' is what Popper or Carnap or, more recently, Kuhn says it is; and that is that. It is to be admitted that some sciences, going through a period of stagnation, now present their results in axiomatic form, or try to reduce them to correlation hypotheses. This does not remove the stagnation, but makes it scientifically respectable. Having no movitation to break out of the circle and much reason (both emotional and financial) to stay in it, philosophers of science can therefore be illiterates with good conscience. Small wonder that intelligent criticism is hard to find . . .[17]

[16] Many papers of positivists and critical rationalists can be summarized in a few lines.
[17] Cf. the criticism of Mach examined in the next chapter.

6

Mach, Einstein and the Popperians

1. In his paper 'Mach, Einstein and Modern Science' in *Brit. J. Phil. Sci.*, 28 (1977), 195ff Elie Zahar states

A. that Einstein, when introducing special relativity, violated 'cardinal' principles of Mach's philosophy (205),[1]

B. that without the violation 'the special theory of relativity would never have seen the light of the day' (195), and he infers

C. that Mach's philosophy 'was largely irrelevant to the development of modern physics' (195).

Now the fact that Einstein violated *some* 'principles' of Mach's philosophy does not show that his ideas were independent of *other* principles of that philosophy; nor does it show that the philosophy did not influence other parts of modern physics, such as the quantum theory: Zahar's arguments do not establish C.

2. They also fail to establish A.

Zahar mentions four conflicts between Mach's and Einstein's ideas and procedures. They are A1: relativity 'is a *causal* theory in the traditional sense of the word' (203) i.e. a theory involving an asymmetry, whereas Mach's functional approach eliminates all asymmetries (202). A2: Einstein's definition of simultaneity violates 'one of Mach's cardinal methodological principles' (205) and it is also 'radically different' (203) from Mach's definition of mass. A3: the special theory of relativity uses inertial frames which, being unobservable, are anathema to Mach. 'It is therefore small wonder' writes Zahar on this point 'that Mach should have disowned relativity' (206). Inertial frames are also needed to make sense of general covariance hence Mach, when arguing for general covariance, 'was violating a tenet of his own branch of positivism' (212). A4: for Einstein (*a*) 'laws have an ontological status over and above the observables which they correlate with one another' (212) and (*b*) they must have such special status, for otherwise the arguments for general covariance again fail to make sense.

[1] According to Zahar the 'cardinal principles' are principles for the definition of scientific concepts, they demand operational definitions, absence of theoretical presuppositions, independence of the theory that contains the concept (202f).

3. But Mach, far from treating inertial systems as metaphysical monsters, examined their properties and watched with interest and approval the attempts to make them a basis for astronomical measurements. He regards corollary 5 of Newton's *Principia*[2] 'as providing the only practically useful (and most likely approximate) inertial system'[3] and he presents L. Lange's 'attractive idea' concerning the definition of such systems.[4] The definition, Mach remarks, separates the 'conventional' and the empirical elements of the law of inertia which consists in a 'restriction of the kinematically possible manifold'. Mach also describes Seeliger's research concerning the relation between inertial systems and the 'empirical coordinate systems actually in use in astronomy'. The attempts to establish 'ephemeris time' and a fundamental reference system[5] which are attempts to make inertial systems the basis of astronomical observations are well known to him and are regarded by him as important empirical investigations: Mach's aversion to relativity has no connection with the occurrence of inertial systems in it.

4. A4(*a*) is also incorrect. In his essay 'Physics and Reality'[6] which criticizes the quantum theory for its '*incomplete* representation of real things' (325f, original italics). Einstein also explains what he means by 'real existence':

> Out of the multitude of our sense experience we take, mentally and arbitrarily, certain repeatedly occurring complexes of sense impressions . . . and correlate to them a concept – the concept of the bodily object. Considered logically this concept is not identical with the totality of sense impressions referred to; but it is a free creation of the human (or animal) mind. On the other hand this concept owes its meaning and its justification exclusively to the totality of the sense impressions which we associate with it.
>
> The second step is to be found in the fact that, in our thinking (which determines our expectations), we attribute to this concept of a bodily object a significance which is to a high degree independent of the sense impressions which originally gave rise to it. This is what we mean when we attribute to the bodily object a 'real existence' [291].

This passage is pure Mach[7] as can be seen from what Mach says about concepts in *Erkenntnis und Irrtum*:[8] according to Einstein the issue in quan-

[2] In *Mathematical Principles of Natural Philosophy*, trans. A. Motte, ed. and rev. F. Cajori (Berkeley, 1960), 20f.

[3] *Die Mechanik*, 9th edition (Leipzig, 1933), xvii.

[4] *Mechanik*, 233. Mach mentions some *technical* difficulties of Lange's definition, but he nowhere regards it as metaphysical.

[5] Cf. the explanations given in the *Explanatory Supplement to the Astronomical Ephemeris and Nautical Almanach* (London, 1961).

[6] Reprinted in *Ideas and Opinions* (New York, 1954), 290ff, original italics.

[7] Except that Mach's account is more realistic than Einstein's: Mach ascribes to a natural development what Einstein regards as a 'free creation of the human (or animal) mind'. For Einstein concepts are freely at our disposal. For Mach their formation obeys psychological laws.

[8] (Leipzig, 1917; first published in 1905.) Both steps are explained already in Mach's earlier writings, e.g. in *Populärwissenschaftliche Vorlesungen* (Leipzig, 1896), 207, 213 (first presented at a meeting of the Vienna Academy in 1882).

tum theory is not an 'ontological' issue (212) whatever *that* means; it is an issue over different modes of 'arbitrarily' ordering sense data.[9]

Confronted with Machian utterances by scientists Zahar speaks (more than once) of 'paying lip service to Machian positivism' (195) and he adds that scientists often do one thing and say another (197): obviously it needs a philosopher to find out what is really going on. But the above quotation cannot be explained away in this simple fashion. It is part of a long article on the situation in physics; it prepares a criticism of the quantum theory; it does not belong to Einstein's early period where some 'lip service' to Mach *might* have occurred (though it is not at all clear why Einstein should pay 'lip service' to any philosophy) but to the 'reactionary' part of Einstein's life when to outsiders he more than ever seemed to be an 'old fashioned realist' (195). The quotation shows that he was 'old fashioned' in the sense of Mach: his objections are conceptual, not 'ontological'. And the article as a whole shows that the saying–doing dichotomy is anything but realistic in Einstein's case: it is very implausible to assume that Einstein, when directing attention to epistemological matters closely connected with the physical research he was engaged in should have failed to know what he was talking about.[10] A look at A4(*b*) shows that he certainly did not on the matter at issue.

5. Zahar says that arguments about covariance involve the behaviour of laws in different reference systems and therefore assume 'that laws and their form possess an objective status' (212); they also assume coordinate systems which are forbidden by Mach and run counter to the principle, allegedly defended by Mach, that the simplest formulation of a law is the

[9] The same applies to Boltzmann's defence of atomism. 'Hertz made it very clear to physicists' Boltzmann said in 1899 (*Populäre Schriften* (Leipzig, 1905), 215) 'what philosophers most likely have said long ago, that a theory cannot be an objective thing, in agreement with nature but can at most be a mental picture of phenomena, being related to them like the sign to the thing signified'. (Hertz had used the term *Scheinbild*: phantom picture.) Boltzmann, accordingly, regarded theories such as the atomic theory as 'mental pictures' (142); his defence of atomism was not an 'ontological' argument, but an argument concerning the superiority of some pictures over others and he emphasized that arguments of this kind were perfectly in agreement with Mach's general philosophy. Mach, on the other hand ('whose relevant writings greatly contributed to the clarification of my own views' Boltzmann, 142, n.1) had no objections to even the most outlandish ones of Maxwell's models and ascribed to them 'Maxwell's extraordinary success': *Erkenntnis und Irrtum*, 229f. These and other matters are explained in detail in M. S. Curd's thesis *Ludwig Boltzmann's Philosophy of Science* (Pittsburgh, 1978) where the 'hagiographic' account of the 'realist' Boltzmann's battle against shortsighted positivists such as Mach and Ostwald is laid to rest once and for all.

[10] Einstein is the last of the late nineteenth-century scientist-philosophers who like Maxwell, Hertz, Mach, Boltzmann and Duhem designed philosophies for advancing science and used (corrected) these philosophies in their research. Today philosophers are outsiders with correspondingly unrealistic views of history and scientific method. Zahar wants to convince us that these outsiders have a better view of science than the most outstanding scientists. His article tells otherwise. Cf. also ns. 47–52.

best: 'there is no good reason, from a Machian viewpoint, why the choice of a certain coordinate system should not prove more convenient than another' (212). The last point is of course true and this is why *particular problems* such as the relativistic advance of the perihelion of Mercury are calculated in special reference systems (Schwarzschild's calculation). But if the purpose is to get laws connecting the special cases then 'physics, for reasons of economy, often gives us equations between functions of the primary variables and not equations between the primary variables themselves'[11] and the task is then to make *these* equations as simple as possible: Zahar has a rather simpleminded view of Mach's conception of economy. But does not the formation of functions of laws assume that laws have an 'objective status'? Yes, but not in Zahar's sense: talking about blueprints assumes that blueprints are 'objective' but not in the sense that they have an 'ontological status *over and above*' the things they are blueprints of. The 'objectivity' of a blueprint, however, can be easily explained in Einstein's (i.e. Mach's) manner, by giving an account of the way in which its elements hang together. Reference systems, finally, are elementary for a Machian (see the refutation of A3 above). This concludes the refutation of A4.

6. A1 and A2 remain. As regards A2 the answer is that Zahar has interpreted a difference between definitions (definition of mass on the one side, definition of simultaneity on the other) as a difference between philosophies (Mach's and Einstein's). But Mach, far from subjecting definitions to rules 1–3 (202f) distinguishes various types of definitions, purely 'nominal stipulative' (207) definitions included. For example, he considers the following definitions of a time metric: (*a*) *t* is proportional to the angle of the earth with respect to some inertial system; (*b*) *t* is proportional to the logarithm of this angle; and he comments on the arbitrary and rival-hypotheses-compatible (207) nature of the stipulations.[12] Einstein's definition of simultaneity falls into this category and is therefore anything but a 'viola[tion] . . . of Mach's cardinal methodological principle' (205). Nor did Mach ever subject definitions to condition 2 (203). Quite the contrary: in his definition of mass he explicitly states the dependence of the definition on 'the existence of a special, acceleration-determining quality of bodies'[13] i.e. on what Zahar calls theoretical presuppositions. Condition 3 is not Machian either. For although Mach objects to what one might call direct circularity (explicit definitions which contain the thing to be defined) he is

[11] 'Die oekonomische Natur der Physikalischen Forschung' quoted from *Populärwissenschaftliche Vorlesungen*, 228.

[12] *Erkenntnis und Irrtum*, 435. Cf. also Mach's comments on the conventional elements of the law of inertia mentioned in the text to n.3 above.

[13] *Mechanik*, 211.

well aware of the fact that concepts, being dependent on each other in many ways, cannot always be explained in isolation. Zahar's condition I, however, (201) can be refuted at once: 'Which concepts we form' writes Mach,[14] 'how we delimit them with respect to each other – all that can be decided only by the needs of practice, or of science'. Mach is much more liberal than modern philosophers of science and his 'cardinal methodological principle' (205) is nothing but a chimaera in Zahar's head.

A1, finally, overlooks the fact that Mach paid great attention to unidirected processes[15] and occasionally regarded them as fundamental: not all connections are symmetrical and those which are 'are secondary, not basic'.[16] Zahar connects Mach's alleged failure to account for asymmetry to his 'conceptual – as opposed to the propositional – approach' (202). He overlooks that Mach's concept are concepts *in use*,[17] i.e. they have propositional content, while the arrow of implication (203), being reducible to disjunction, contains no asymmetry. What is needed in matters of symmetry versus asymmetry is not an emancipation from an alleged 'conceptual' philosophy and its replacement by a Popperian propositionalism;[18] it is 'more research' concerning natural processes.[19]

7. There are many other mistakes. Zahar: 'Mach thought . . . he could boil everything down to spatial measurements' (206). Mach: 'there is an unchangeable physical time object just as there is an unchangeable physical space object – the rigid body.'[20] Zahar: '*Elements* . . . are the simplest component parts of sensations' (200, original italics); Mach: 'elements are sensations *only insofar*' as we consider their dependence on a particular complex of elements, viz. our body; '*they are simultaneously physical objects*, viz. insofar as we consider other functional dependencies'.[21] If we disregard the dependencies, elements are therefore as much physical objects as they are sensations.[22] Zahar's biggest mistake which clearly shows his ignorance concerning Mach's philosophy is to describe it as 'strictly positivistic' (198).[23] The title of Mach's first major work, alone, should have given him

[14] *Erkenntnis und Irrtum*, 129.

[15] *Mechanik*, 218; *Analyse der Empfindungen* (Jena, 1922), 286f, esp. 286, lines 12ff.

[16] *Erkenntnis und Irrtum*, 443.

[17] 126ff and passim: Mach has a decidedly constructivist (in the sense of mathematical constructivism) attitude towards concepts.

[18] Zahar, like all good Popperians, ascribes even the most trite and useless idea to Popper. In the case of propositionalism the origin is clear: it is *Aristotle*.

[19] *Mechanik*, 219. [20] *Erkenntnis und Irrtum*, 393. [21] *Analyse*, 13, my italics.

[22] Mach seems inclined to count some of Duhem's 'qualities' among the elements. Cf. his introduction to the German translation of Duhem's main philosophical work, *Ziel und Struktur der Physikalischen Theorien* (Leipzig, 1908). But 'qualities', for Duhem, are physical primitives such as currents, charges and so on.

[23] R. Carnap, *Der Logische Aufbau der Welt* (Hamburg, 1974; first published 1928), 87, notices with surprise that Mach's 'basis does not consist of events belonging to the perceiving subject' ('die nichteigenpsychische Basis') and points out that this feature 'constitutes a

cause for doubt. An author who writes an *Analysis* of sensations obviously does not take sensations as basic. And, indeed, Mach insists on an analysis of sensations including an examination of their dependence on physiological conditions;[24] he regards results of such an analysis as 'temporary' (*Analyse*, 24) and sensation talk as involving a 'one sided theory' (18); he asserts that introspection does not suffice as an instrument of analysis but must be combined with a physiological study;[25] he points out that our ideas affect not only 'the development of *experience*' but 'enrich every . . . fact' (*Erkenntnis und Irrtum*, 245); he objects to Mill's aversion to any hypothesis which 'favours what has been found over what has still to be discovered' (249); he encourages the researcher 'not to be too fearsome when introducing hypotheses. Quite the contrary, a little boldness in these matters furthers research' (247); he quotes with approval Priestley's observation that 'very lame and imperfect theories are sufficient to suggest useful experiments while serving to correct those theories and give birth to others more perfect. These then occasion further experiments *which bring us still nearer to the truth*, and in this method of approximation we must be content to proceed . . .' (240f, my italics). He describes himself as a 'non empiricist' or 'not-only-empiricist'[26] and, considering his reconstruction scheme of science in terms of hypothetical elements (*not* sensations) he warns that 'there is no need to turn *this passing point of view* into a system for life whose slaves we then become . . .'.[27] There is a great distance indeed between Mach's 'philosophy'[28] and Zahar's rather unbelievable distortion of it. To ascribe such views to Mach shows not only an astonishing ignorance of relevant historical material; to assume that a scientist-philosopher and social critic of Mach's calibre[29] should in the end have come up with such a

disturbing element in the total picture of his (Mach's) ideas'. Obviously he has formed a false conception of this picture, and he now notices difficulties. No such difficulties exist for the Popperian Zahar.

[24] Cf. text to n.32 below. [25] Cf. text to n.32 below.

[26] Hugo Dingler, *Die Grundgedanken der Machschen Philosophie* (Leipzig, 1924), 61n. Cf. also vol. 2 of the notebooks, 16 pages from the back (contained in Dingler's book) where Mach puts sensations in their place and wants to free science from their dominance: Mach criticized not only realism, but *all* one sided philosophies. Zahar refers to Dingler's book, but seems to have read only a tiny part of it.

[27] *Populärwissenschaftliche Vorlesungen*, 226. One should compare this quotation with the Einstein quotation in the text to n.47.

[28] Mach refused to regard his hints or 'apercus' (*Analyse*, 39) as a philosophy: 'I am a scientist, not a philosopher . . . and I think I am innocent of the fact that people have regarded me as an idealist, Berkeleyan, even a materialist' (*Analyse*, 39). 'Above all there is no Machian philosophy' he writes in the introduction to *Erkenntnis und Irrtum*, vii, 'there is, at most, a methodology of science and a psychology of knowledge *and like all scientific theories these things must be regarded as preliminary and incomplete attempts*' (my italics). There are many philosophers with whom Mach feels he has 'points of contact', Spinoza among them (*Analyse*, 38). Note how much more dogmatic critical rationalists sound when pronouncing on their 'philosophy'.

[29] 'I don't think anyone ever gave me so strong an impression of pure intellectual genius. He apparently has read everything and thought about everything, and has an absolute sim-

sophomoric theory shows in addition an enormous lack of imagination and historical understanding.[30]

8. There is, then, *no conflict* between Mach's philosophy and the (genesis of the) special theory of relativity. But is there a positive relation? Are there elements in Einstein's early research that show a similarity to Mach's philosophy and may therefore be regarded as evidence of Machian influences? On the other hand, what are we to make of Mach's criticism of relativity? I shall now give some very tentative answers to these questions.

To explain Mach's criticism I draw attention to two passages contained in a longer quotation from Mach by Zahar (197). These passages both refute Zahar and also point to an answer fairly different from Zahar's. The fact that Zahar quotes them and yet continues on his own life of attack shows to what an extent Popperian mythology is capable of blinding its followers.

Mach discredits relativity because he finds it 'growing more and more dogmatical' (first reason) and because of 'particular reasons', 'based on the physiology of the senses, the theoretical ideas and, above all, the conceptions resulting from experiments' (second reason).

I connect the second reason with Mach's attempt to find a point of view that is not restricted to a particular subject such as physics, or psychology 'but can be retained in all domains' (*Analyse*, 255). Or, to express it differently, but still in agreement with Mach's intentions (*Analyse*, 26): he sets himself the task of finding theories that cut across traditional boundaries such as the boundary between physics and psychology without being beset by the mind–body problem. What have 'experiments' got to do with the construction of such a theory?[31] In an earlier essay Mach comments that 'the mental field', i.e. the domain where thoughts, emotions, etc. appear and disappear 'can never be fully explored by introspection. But introspection combined with physiological research which explores the physical connections can put this field clearly before us and only thereby make us acquainted with our inner being':[32] introspection does not suffice; it grasps only part of 'mental events'; their whole nature is revealed by introspection plus physiology. The partly physical character of the research leads to the expectation that our 'inner being' *may without break merge* into the physical

plicity of manner and winningness of smile when his face lights up, that are charming . . .' William James about Ernst Mach, in a letter to his wife of 2 November 1882 quoted from Joachim Thiele, *Wissenschaftliche Kommunikation* (Kastellaun, 1978), 169. Of course, it is difficult for a dweller in the slums of late twentieth-century philosophy of science to guess what an intelligent man may or may not assume.

[30] Or else it shows that Popperians, being engulfed by the banality of their own 'philosophy', have completely lost their sense of perspective.

[31] Note that Zahar cannot explain the role of *experiments* in Mach's objection: one does not need experiments to show that a theory is metaphysical.

[32] *Populärwissenschaftliche Vorlesungen*, 228.

world thus providing a unified account of 'mental' and 'physical' events and laws. A purely physical theory of space (or spacetime) does not fit into such a research programme. This is the 'particular reason'.[33]

Now despite this disadvantage the theory of relativity might still be welcomed as a step in the right direction were it not for the fact that it had been 'growing more and more dogmatical' (first reason). Was this the case? Was there really an increase of dogmatism, or what Mach regarded as dogmatism in physics, in connection with the theory of relativity? Not on Einstein's part but certainly on the part of Planck, Mach's arch enemy.[34] Planck, after some wavering, defended a reality of the kind Mach and Boltzmann had rejected; he regarded relativity as a step towards this reality, made himself many contributions towards the development and extension of the theory[35] and so furthered its dogmatic interpretation. I suggest that this explains Mach's criticism and the harshness of it but it also shows that *the criticism does not touch Einstein's achievements*. It concerns only their misuse for dogmatic philosophical purposes.[36]

9. Is Einstein's work influenced by Mach? I think it is. However the influence becomes clear only if one disregards technical details and looks at the wider picture. We notice then that Einstein strives for the most direct

[33] There are various hypotheses concerning the 'experiments'. Thus Herneck conjectures that Mach objected to the assumption of the constancy of c and that the 'experiments' dealt mainly with these matters: D. K. Heller, *Ernst Mach* (New York, 1964), 142. According to Herneck, Philipp Frank gave Dingler part of the responsibility for Mach's attitude (*Phys. Blätter*, 17 (1961), 276). However, the fact that Mach emphasized the need to combine psychological, physiological, physical as well as biological investigations both in his early and in his late writings and that he deplored a purely physical approach to the problem of space seems to favour my interpretation: relativity was acceptable especially as it introduced the interdependence of spatial and temporal aspects (cf. n.46) but it did not go far enough and it became a hindrance to research if interpreted dogmatically, in the manner of Planck.

[34] Cf. the rather acrimonious exchange between Mach and Planck, reprinted in *Physical Reality*, ed. S. Toulmin (New York, 1968). Although the tone of learned exchange was at the time a little less vapid than it is today, this was the only polemic Mach ever published.

[35] Cf. the account and the literature in Stanley Goldberg 'Max Planck's Philosophy of Nature and his Elaboration of the Special Theory of Relativity' *Historical Studies in the Physical Sciences* 7 (1976), 78–125.

[36] Similar remarks apply to Mach's rejection of atoms. First a little myth should be laid to rest. Stefan Meyer of the Institute for Radium Research at the University of Vienna reports, and many writers have repeated, that Mach, having been shown a spinthariscope said 'jetzt glaube ich an die Atome' *Sitz. Ber. Oesterr. Akad. Wiss. Math. Naturwiss. Kl.* ser. 2a, vol. 159, no. 1/2, 5. Heinz Post has pointed out that Mach was a kind person, that Meyer was very enthusiastic and rather deaf and that he most likely misunderstood the kind remarks Mach made in such a personally difficult situation. And who would think that Mach, who had not been impressed by some very powerful arguments (e.g. Brownian motion), should fall for such a flimsy exhibition? Secondly, Mach's objections were firmly grounded in some very real difficulties of the atomic theory which were often forgotten in view of apparent successes. Finally, the nineteenth-century atom has indeed disappeared and has been replaced by a much more complex entity, constituted out of relations between well-defined measurements and no longer in conflict with Mach's ideas.

expression of phenomena: electric induction depends only on the relative motions of the systems concerned while the ether theories treat the effects of the motion of a magnet (relative to the ether) differently from the effects of the motion of a closed electric circuit.[37] Lorentz explains null results approximately and using a variety of rather doubtful hypotheses[38] while Einstein regards them as basic facts.[39] In his investigations of the structure of radiation Einstein does not start from basic theory, as Planck did, but from uncomprehended but empirically adequate laws and develops their 'ontological' consequences. 'The key to his reasoning' writes Martin Klein on this procedure[40] 'was his reversal of Planck's procedure. Instead of trying to derive the distribution law from some more fundamental starting point, he turned the argument around. Planck's law had the solid backing of experiment; why not assume its correctness and see what conclusion it implied as to the structure of radiation?' He plays 'theoretical pictures'[41] (his term, most likely taken from Boltzmann whom he read carefully)[42] off against each other to find their limits.[43] Like Mach, Einstein is mainly interested in 'theories of principle'[44] i.e. in theories which cover a wide variety of phenomena without assumptions about details of their constitution (this also establishes a link with Duhem). Einstein's remarks on geometry are in perfect agreement with what is found in the relevant chapters of *Erkenntnis und Irrtum*[45] and in earlier writings.[46] As regards

[37] 'On the electrodynamics of Moving Bodies' quoted from *The Principle of Relativity* (New York, 1921), 37.

[38] Cf. the articles by Lorentz in *The Principle of Relativity*.

[39] 'I have not been able to obtain for the equations referred to moving axes *exactly* the same form as for those which apply to a stationary system.' H. A. Lorentz, *The Theory of Electrons* (New York, 1952), 230. 'Einstein simply postulates what we have deduced, with some difficulty, and not altogether satisfactorily, from the fundamental equations of the electromagnetic field. By doing so, he may certainly take credit for making us see in the negative result of experiments like those of Michaelson, Raleigh and Bruce, not a fortuitous compensation of opposing effects, but the manifestation of a general and fundamental principle.' Actually, Einstein did not 'make us see it' that way, for there was no explanation, he simply took this situation as his starting point: 'Wir wollen diese Vermutung zur Voraussetzung erheben' *Das Relativitaetsprinzip* (Leipzig, 1923), 26; English translation, *Principle of Relativity*, 38.

[40] 'Einstein and the Wave–Particle Duality' in *The Natural Philosopher* (New York, 1964) III, 9. Cf. also T. S. Kuhn, *Black Body Theory and the Quantum Discontinuity 1894–1912* (New York, 1978), 180f.

[41] Cf. the first page of his paper on the photoelectric effect reprinted in D. ter Haar, *The Old Quantum Theory* (New York, 1967), 91.

[42] Cf. the references in Seelig's biography. For Boltzmann, cf. n.9 above.

[43] Mach was well aware of the fact that different theories could cover some domains equally well. Cf. also his remarks on analogues in science *Erkenntnis und Irrtum*, 220ff.

[44] Cf. the autobiographical notes in *Albert Einstein: Philosopher-Scientist*, ed. P. A. Schilpp (New York, 1951), esp. 53.

[45] Cf. ch. 1 of Einstein's, *The Meaning of Relativity* (Princeton, 1922) with *Erkenntnis und Irrtum*, 353ff.

[46] Cf. Mach's essay on the development of our ideas of space in *Z. Phil. Phil. Kritik*, New ser. 49 (1866), esp. 232. Cf. also Mach's note concerning Minkowski's 'Raum und Zeit' in the new (1909) edition of 'Die Geschichte und die Wurzel des Satzes von der Erhaltung der Arbeit'

method Einstein emphasizes the great difference between the work of the scientist and the work of the epistemologist (philosopher of science): 'The external conditions' writes he[47] 'which are set [for the scientist] by the facts of experience do not permit him to let himself to be too much restricted, in the construction of his conceptual world, by the adherence to an epistemological system. He, therefore, must appear to the systematic epistemologist as a type of unscrupulous opportunist.' 'Yes, I may have started it' he tells Infeld about a new fashion in physics, 'but I regarded these ideas as temporary. I never thought that others would take them so much more seriously than I did.'[48] Or, in a lighter mood: 'A good joke should not be repeated too often.'[49] Compare this with Mach's 'the schematisms of *formal* logic and also of *inductive* logic are not of much use, for the intellectual situations are never repeated in exactly the same way. But the examples of great scientists are very instructive . . .',[50] the already quoted passage 'there is no need to turn *this passing point of view* (i.e. Mach's own suggestions) into a system for life whose slaves we then become'[51] as well as with Mach's modest and somewhat mocking attitude towards his own 'philosophy'[52] and you will see that Einstein and, as a matter of fact, all creative scientists are much closer to Mach than to, say, Popper. Mach, Einstein, Bohr use philosophy as an instrument of research that guides research but is also transformed by it, while philosophers pontificate on a process they do not understand and would not know how to change. Small wonder they want to persuade us that Mach's ideas are as vapid as their own.

with reference to the relevant chapters of *Erkenntnis und Irrtum*: 'Space and time are not regarded as self contained entities but as forms of dependence of the phenomena. I therefore move towards the principle of relativity which is also asserted (festgehalten) in "Mechanik" and "Wärmelehre"' (60).

[47] *Albert Einstein*, 683f.

[48] Quoted from R. W. Clark, *Einstein* (New York, 1971), 340.

[49] Philipp Frank, *Einstein, His Life and Times* (London, 1948), 261.

[50] *Erkenntnis und Irrtum*, 200. The same applies of course to the various theories of rationality which are of no use to the scientist but only provide philosophers with material for endless quarrels. 'Die Wissenschaft ist kein Advokatenkunststueck' says Mach (*Erkenntnis und Irrtum*, 402, n.1, against the Euclidian mode of discussion); and he is certainly right.

[51] Cf. n.27 above.

[52] Cf. n.28 above.

7
Wittgenstein's *Philosophical Investigations*

In discussing this book[1] I shall proceed in the following way: I shall first state a philosophical theory T, which is attacked throughout the book. In doing so I shall not use the usual statement of the theory (if there is any) but Wittgenstein's, which may, of course, be an idealization. Secondly, I shall show how the theory is criticized by Wittgenstein: first, using an example (which plays a considerable role in his arguments, but which I have used to present other arguments as well), then discussing in general terms the difficulties revealed by the example. Thirdly, I shall state what seems to be Wittgenstein's own position on the issue. This position will be formulated as a philosophical theory, T', without implying that Wittgenstein intended to develop a philosophical theory (he did not). Finally I shall discuss the relation between the theory stated and Wittgenstein's views on philosophy and I shall end up with a few critical remarks.[2]

For brevity's sake I shall introduce three different types of quotation marks: The usual quotation marks ("...") enclosing Wittgenstein's own words, daggers († . . . †) enclosing further developments of his ideas and general remarks, asterisks (* . . . *), enclosing critical remarks. Text without any of these quotation marks is an abbreviated statement of what Wittgenstein is saying.

1. †The theory criticized is closely related to medieval realism (about universals) and to what has recently been termed "essentialism". As presented by Wittgenstein, it includes the following five main items.

†(1) "Every word has a meaning. This meaning is correlated with the word. It is the object, for which the word stands" (1; 90, 120).[3] Meanings exist independently of whether or not any language is used and which language is used. They are definite single objects and their order "must be *utterly simple*" (97).

[1] This paper was written in 1952 in German and translated into English by G. E. M. Anscombe.

[2] Although many different problems are discussed in the *Investigations*, it seems to me that the criticism of T (or the assertion of T') is to be regarded as the core of the book. I shall therefore concentrate on elaborating T and T', and I shall omit all other problems (if there are any).

[3] Parenthetical references are to the numbered sections of part I of the *Philosophical Investigations*, unless otherwise indicated.

†(2) As compared with this definiteness and purity of meanings (their order "must ... be of the purest crystal" (97)), "the actual use ... seems something muddied" (426). This indicates an imperfection of our language.

†(3) This imperfection gives rise to two different philosophical problems: (*a*) The philosopher has to find out what a word '*W*' stands for, or, as it is something expressed, he has to discover the *essence* of the object that is designated by '*W*', when its use in everyday language is taken into account. From the knowledge of the essence of *W* the knowledge of the whole use of '*W*' will follow (264, 362, 449). (*b*) He has to build an ideal language whose elements are related to the essences in a simple way. The method of finding a solution to problem (*a*) is analysis. This analysis proceeds from the assumption that "*the essence is hidden from us*" (92) but that it nevertheless "'*must*' be found in reality" (101). However different the methods of analysis may be – analysis of the linguistic usage of '*W*'; phenomenological analysis of *W* ('deepening' of the phenomenon *W*); intellectual intuition of the essence of *W* – the answer to problem (*a*) "is to be given once for all; and independently of any future experience" (92). The form of this answer is the definition. The definition explains why '*W*' is used in the way it is and why *W* behaves as it does (75; 97, 428, 654). The solution of (*b*) is presupposed in the solution of (*a*); for it provides us with the terms in which the definitions that constitute the solution of (*a*) are to be framed. A definite solution of (*b*) implies a certain form of problem (*a*). If it is assumed, for example, that sentences are word-pictures of facts (291; cf. *Tractatus Logico-Philosophicus* 2.1; 4.04) then 'What is a question?' is to be translated into 'What kind of fact is described by a question? The fact that somebody wants to know whether ..., or the fact that somebody is doubtful as to ..., etc.?'

†(4) Asking how the correctness of a certain analysis may be checked, we get the answer that the essence can be *experienced.* This experience consists in the presence of a mental picture, a sensation, a phenomenon, a feeling, or an inner process of a more ethereal kind (305). 'To grasp the meaning' means the same as 'to have a picture before one's inner eye' and "to have understood the explanation means to have in one's mind an idea of the thing explained, and that is a sample or a picture" (73). The essence of the object denoted, the meaning of the denoting expression (these are one and the same thing; cf. 371, 373) follows from an analysis of this picture, of this sensation; it follows from the exhibition of the process in question (thus the essence of sensation follows from an analysis of my present headache (314)). It is the presence of the picture which gives meaning to our words (511, 592), which forces upon us the right use of the word (73, 140, 305, 322, 426, 449), and which enables us to perform correctly an activity (reading, calculating) the essence of which it constitutes (179, 175, 186, 232). Under-

standing, calculating, thinking, reading, hoping, desiring are, therefore, mental processes.

†(5) From all this it follows that teaching a language means showing the connection between words and meanings (362) and that "learning a language consists in giving names to objects" (253). So far we have the description of *T*, as it is implicitly contained in the *Philosophical Investigations*.

2. †In criticizing *T*, Wittgenstein analyses *T*(4) and in this way shows the impossibility of the program *T*(3) as well as the insolubility of the problems connected with this program. This implies that, within *T*, we shall never be able to know what a certain word '*W*' means or whether it has any meaning at all, although we are constantly using that word and although the question of how it is to be used does not arise when we are not engaged in philosophical investigations. But did not this paradox arise because we assumed that meanings are objects of a certain kind and that a word is meaningful if and only if it stands for one of those objects; i.e. because we assumed *T*(1), (2) to be true? If, on the other hand, we want to abandon *T*(1), (2), we meet another difficulty: words have, then, no fixed meaning (79). "But what becomes of logic now? Its rigour seems to be giving way here. – But in that case doesn't logic altogether disappear? – For how can it lose its rigour? Of course not by our bargaining any of its rigour out of it. – The *preconceived idea* of crystalline purity can only be removed by turning our whole examination round" (108); i.e. by changing from *T* to *T'*. It will turn out that this change cannot be described simply as the change from one *theory* to another, although we shall first introduce *T'* as a new theory of meaning.

†Before doing so we have to present Wittgenstein's criticism of *T*. This criticism is spread throughout the book. It consists of careful analyses of many special cases, the connection between which is not easily apprehended. I have tried to use *one* example instead of many, and to present as many arguments as possible by looking at this example from as many sides as possible. All the arguments are Wittgenstein's; some of the applications to the example in question are mine.

3. †The philosopher is a man who wants to discover the meanings of the expressions of a language or the essences of the things designated by those expressions. Let us see how he proceeds. Let us take, for example, the word 'reading'. "Reading is here the activity of rendering out loud what is written or printed; and also of writing from dictation, writing out something printed, playing from a score and so on" (156).

†(A) According to *T*(1) we have to assume that the word 'reading' stands for a single object. Now, there is a variety of manifestations of reading: reading the morning paper; reading in order to discover misprints (here one

reads slowly, as a beginner would read); reading a paper written in a foreign language that one cannot understand but has learned to pronounce; reading a paper in order to judge the style of the author; reading shorthand; reading *Principia Mathematica*; reading Hebrew sentences (from right to left); reading a score in order to study a part one has to sing; reading a score in order to find out something about the inventiveness of the composer, or to find out how far the composer may have been influenced by other contemporary musicians; reading a score in order to find out whether the understanding of the score is connected with acoustic images or with optical images (which might be a very interesting psychologial problem). But this variety, without "any one feature that occurs in all cases of reading" (168), is only a superficial aspect. All these manifestations have something *in common* and it is this common property which makes them manifestations of *reading*. It is also this property that is the essence of reading. The other properties, varying from one manifestation to the other, are accidental. In order to discover the essence we have to strip off the particular coverings which make the various manifestations *different* cases of reading. But in doing so (the reader ought to try for himself!) we find, not that what is essential to reading is hidden beneath the surface of the single case, but that this alleged surface is one case out of a family of cases of reading (164).[†]

> Consider for example the proceedings which we call 'games'. I mean board-games, card-games, ball-games, Olympic games and so on. What is common to them all? – Don't say: 'There *must* be something common or they would not be called "games" '; but *look and see* whether there is anything common to *all* – for if you look at them you will not see something that is in common to all, but similarities, relationships and a whole series of them at that . . . And the result of this examination is: we see a complicated network of similarities overlapping and criss-crossing . . . I can think of no better expression to characterize these similarities than 'family-resemblances'; for the various resemblances between members of a family: build, features, colours of eyes, gait, temperament, etc., etc., overlap and criss-cross in the same way. – And I shall say: 'games' form a family [66f].

> And in the same way we also use the word 'reading' for a family of cases. And in different circumstances we apply different criteria for a person's reading [164].

[†](B) Looking at the outer manifestations of reading we could not discover the structure suggested by $T(1)$. Instead of an accidental variety centring in a well-defined core we found "a complicated network of similarities" (66). Does that fact refute $T(1)$? Surely not; for if a philosopher wants to defend $T(1)$, there are many possible ways of doing so. He may admit that the *overt behaviour* of the person reading does not disclose any well-defined centre, but he may add that reading is a *physiological process* of a certain kind. Let us call

this process the reading process (R P). Person P is reading if and only if the R P is going on within (the brain or the nervous system of) P. (Cf. 158.) But the difficulties of this assumption are clear. Consider the case of a person who does not look at any printed paper, who is walking up and down, looking out of the window and behaving as if he were expecting somebody to come; but the R P is going on within his brain. Should we take the presence of the reading process as a sufficient criterion for the person's reading, adding perhaps that we had discovered a hitherto unknown case of reading? (Cf. 160.) It is clear that in a case like that we should, rather, alter some physiological hypotheses. If, again, reading is a physiological process, then it certainly makes sense to say that P read 'ali' within 'totalitarianism' but did not read before he uttered those sounds and did not read afterward either, although anybody who observed the outer behaviour of P would be inclined to say that P had been reading the whole time. For it is quite possible that the R P should be present only when P is uttering 'ali' (cf. 157). It seems, however, that it is quite meaningless to hypothesize that in the circumstances described a person was reading only for one second or two, so that his uttering of sounds in the presence of printed paper before or after that period must not be called 'reading'.

†(C) To the failure of attempts (A) and (B) to discover the essence of reading certain philosophers will answer in the following way: Certainly; that was to be expected.† For reading is a *mental process*, and "the one real criterion for anybody's *reading* is the conscious act of reading, the act of reading the sounds off from the letters. 'A man surely knows whether he is reading or only pretending to read' " (159). †The idea alluded to is this: Just as the sensation *red* is present when we are looking at a red object, so a specific mental process, the reading process (M R P), is present in the mind when we are reading. The M R P is the object of our analysis of reading, its presence makes our overt behaviour a manifestation of reading (etc., as already indicated in *T*(4)). In short, it is thought that this mental process will enable us to solve problems which we could not solve when considering material processes only: "When our language suggests a body and there is none; there, we should like to say, is a *spirit*" (36). But it will turn out that mental processes are subject to the same kind of criticism as material processes: that neither a material nor a spiritual mechanism enables us to explain how it is that words are meaningful and that their meanings can be known; that in pointing to mental processes we cling to the same scheme of explanation as in the physiological or the behaviouristic theory of meaning (considered in the two last sections) without realizing that we are doing so.[4] This can be shown by very simple means: consider the case of a person who does not look at any printed paper, who is walking up and down, looking

[4] This point is elaborated in some detail in G. Ryle's *Concept of Mind* (London, 1949), which should not, however, be taken to agree completely with Wittgenstein's ideas.

out of the window, and behaving as if he were expecting somebody to come; but the MRP is going on in his mind (in his consciousness). Should we take the presence of this mental process as a sufficient criterion for the person's reading, adding, perhaps, that we had discovered a hitherto unknown case of reading? It is clear that we should alter, rather, some psychological hypotheses (the hypothesis that reading is always correlated with the MRP). But the last argument is a simple transformation of the first argument of section (B) with 'MRP' (the mental process which is supposed to be the essence of reading) substituted for 'RP' (the physiological process, which was supposed to be the essence of reading in section (B)). By this substitution the second argument can be used for the present purpose as well.

†(a) Let us now turn to a more detailed investigation of the matter. Let us first ask whether really *every act of reading is accompanied by the MRP*. A few minutes ago I was reading the newspaper. Do I remember any particular mental process which was present all the time I was reading? I remember that I was expecting a friend (actually I looked at my watch several times) and that I was angry because he did not come, although he had promised to do so. I also remember having thought of an excellent performance of *Don Giovanni* which I had seen a few days ago and which had impressed me very much. Then I found a funny misprint and was amused. I also considered whether the milk which I had put on the fire was already boiling, etc. Nevertheless, I was *reading* all the time, and it is quite certain that I was (cf. 171).† "But now notice this: While I am [reading] everything is quite simple. I notice nothing *special*; but afterward, when I ask myself what it was that happened, it seems to have been something indescribable. *Afterward* no description satisfies me. It is as if I couldn't believe that I merely looked, made such and such a face and uttered words. But don't I *remember* anything else? No" (cf. 175; "being guided" instead of 'reading'). †The same applies to activities such as calculating, drawing a picture, copying a blueprint, etc. I *know* of course that I was reading, but that shows only that my knowledge is not based on the memory of a certain sensation, impression, or the like – because there was no such impression.† Compare now another example: Look at the mark ə and let a sound occur to you as you do so; utter it – let us assume it is the sound 'u'. †Now read the sentence 'Diana is a beautiful girl.' Was it in a different way that the perception of the 'eau' (in 'beautiful') led to the utterance of the sound 'u' in the second case? Of course there was a difference! For I *read* the second sentence whereas I did not read when I uttered the 'u' in the presence of the ə. But is this difference a difference of mental content, i.e. am I able to discover a specific sensation, impression, or the like which was present in the second case, and missing in the first case, whose presence made the second case a case of *reading*?† Of course, there were many differences: In the first case "I had told

myself beforehand that I was to let a sound occur to me; there was a certain tension present before the sound came. And I did not say 'u' automatically as I do when I look at the letter U. Further that mark [the ə] was not *familiar* to me in the way the letters of the alphabet are. I looked at it rather intently and with a certain interest in its shape" (166). But imagine now a person who has the feeling described above in the presence of a normal English text, composed of ordinary letters. Being invited to read, he thinks that he is supposed to utter sounds just as they occur to him – one sound for each letter – and he nevertheless utters all the sounds a normal person would utter when reading the text. "Should we say in such a case that he was not really reading the passage? Should we here allow his sensations to count as the criterion for his reading or not reading?" (160). From the negative answer to this question we have to conclude that, even if we were able to discover a difference between the way in which the perception of the ə leads to the utterance of the sound 'u' and the way in which, for example, the perception of the 'eau' within 'beautiful' leads to the utterance of the 'u', this difference – if it is a difference of mental content, of behaviour, etc. – cannot be interpreted as justifying the assumption of an essential difference between cases of reading and not reading.[5]

(b) It may be objected to this analysis that the MRP is sometimes present quite distinctly. "Read a page of print and you can see that something special is going on, something highly characteristic" (165). This is true especially where "we make a point of reading slowly – perhaps in order to see what does happen if we read" (170). Thus one could be inclined to say that the MRP is a subconscious process which accompanies *every* case of reading but which can be brought to light only by a special effort.[6]

[5] There are cases of mental illness where the patient talks correctly although with the feeling that somebody is making up the words for him. This is rightly regarded as a case of mental illness and not, as the adherents of the mental-picture theory of meaning would be inclined to say, as a case of inspiration: for one judges from the fact that the person in questions *talks correctly*, although with queer sensations. Following Locke, a distinction is usually made between impressions of sensation and impressions of reflection. When Wittgenstein talks of sensations, of feelings, of a 'picture in the mind" he seems to mean both. So his investigations are directed against a primitive psychologism (concepts are combinations of impressions of sensation) as well as against a more advanced psychologism (concepts are combinations of impressions of reflection). They are also directed against a presentational realism (concepts are objects of a certain kind, but *having* a concept, or *using* a concept is the same as having an idea in one's mind; i.e. although concepts are not psychological events, their representations in people are), against a theory which Wittgenstein elsewhere described as implying that "logic is the physics of the intellectual realm".

[6] A psychologist or an adherent of the phenomenological method in psychology would be inclined to judge the situation in this way. His intention would be to create a kind of "pure situation" in which a special process comes out quite distinctly. It is then supposed that this process is hidden in every ordinary situation (which is not pure, but) which resembles the pure situation to a certain extent. In the case of reading the pure situation would be: reading plus introspecting in order to find out what is going on. The ordinary situation is: simply reading.

Answer: (1) Reading with the intention of finding out what happens when we are reading is a special case of reading and as such is different from ordinary reading (cf. 170). Nevertheless reading without this intention is also a case of reading, which shows that the reason for calling it a case of *reading* cannot be the presence of a sensation which – admittedly – is present only in special cases and not in the case discussed. Finally, the description of the MRP cannot be a description of reading in general, for the ordinary case is omitted. We should not be misled by the picture which suggests "that this phenomenon comes in sight 'on close inspection'. If I am supposed to describe how an object looks from far off, I don't make the description more accurate by saying what can be noticed about the object on closer inspection" (171).

†(2) Not every kind of introspection is judged in the same way. It is possible that a person who is supposed to find the MRP by introspection, being tired, will experience and describe quite unusual things while thinking all the time that the task which was set him by the psychologist is being performed by giving these descriptions.[7] No psychologist will welcome such a result. Instead of thinking that new and illuminating facts about reading have been discovered, he will doubt the reliability of the guinea pig. From this we have to conclude once more that the sensations experienced in connection with reading, and even those experienced as the essence of reading by the readers themselves, have nothing whatever to do with the question of what reading really is.

†(3) Let us now assume that a reliable observer whom we ask to read attentively and to tell us what happens while he is reading provides us with the following report: 'The utterance is *connected* with seeing the signs; it is as if I were *guided* by the perception of the letters, etc.' (cf. 169, 170, 171). Does he, when answering our question in this way, describe a mental content, as a person who is seeing red and who tells us that he is seeing red describes a mental content: Does he say 'I am being guided by the letters' because the mental content *being guided* is present? Then one would have to conclude that every case of being guided is accompanied by *being guided*, as we assumed at the beginning of section (C) that every case of reading is accompanied by the MRP. But this last assumption has already been refuted, and the other, being completely analogous to it, can be refuted by the same arguments. We have to conclude, therefore, that the possibility of describing the process of reading as a case of being guided does not imply that reading is a mental process, because being guided is not one (cf. 172).[8]

[7] An illustrative example for experiences of this kind may be found in B. Russell, *History of Western Philosophy* (New York, 1945), 145.

[8] The idea that reading is a single object (in spite of the variety of manifestations demonstrated in section (A)) is apparently supported by the fact that one can give a definition like the one we gave at the beginning of section 3 or that one can say that reading is a form of being guided. But let us not be misled by words. For the definition of reading in terms of being

†(c) As already indicated, people usually try to escape from argument (Ca) by assuming that the MRP is a subconscious sensation which has to be brought to light by introspection. A different form of the same escape is the following one: The arguments that have been brought forward so far assume that reading and the MRP can be separated from one another. This, however, is not the case: reading is inseparably connected with the MRP. What occurs separably from reading is not the MRP, but only an erroneous interpretation of something as reading. But how are we to decide whether the MRP itself is present or only something else erroneously interpreted as reading; or, what comes to the same thing, how are we to decide whether we are reading or only believing that we are reading? The given content of consciousness cannot be used for deciding this question, for it is *its* reliability which is to be ascertained. The only possible alternative is to call a sensation a case of the MRP if and only if it is accompanied by reading. But now we assume, contrary to our previous assumption, that we do possess a criterion for reading other than a sensation.

†Another argument against the assumption of a hidden mental content, which may be brought to daylight by introspection or some other mental act, consists in developing the paradoxical consequences of such a view: "How can the process of [reading] have been hidden when I said 'now I am [reading]' *because* I was [reading]?! And if I say it is hidden – then how do I know what I have to look for?" (153; "understanding" replaced by "reading".)[9]

†(d) So far we have shown (by a kind of empirical investigation into the use of the word 'reading') that there is not a mental content which is *always* present when a person is reading, and that therefore giving the criterion for a person's reading cannot consist in pointing out a particular mental content. Now we shall show that even if there were a mental content which is present if and only if a person is reading, we could not take this content to be the essence of reading. Let us assume that a mental content is the essence

guided or the like supports the idea that reading is a single object only if being guided can itself be shown to be a single object. But an analysis similar to the one sketched in section (A) will show that this is not the case.

One of the main reasons for the wide acceptance of the assumption that it is possible to discover the essence of reading by introspection is the fact that the great number of manifestations of reading is usually not taken into account. Beset by theory *T* we *think* (173, 66) that acute observation must disclose the essence and that what we find in acute observation is hidden in the ordinary case of reading (*T*(4)). But our knowledge of the ordinary case is much too sketchy to justify that assumption "A main cause of philosophical disease – a one-sided diet: one nourishes one's thinking with only one kind of example" (593).

[9] The same criticism applies to the method of the phenomenologists. How do they know which phenomenon is the 'right' one? They proceed from the assumption that the essence is not open to general inspection but must be discovered by some kind of analysis which proceeds from an everyday appearance. In the course of this analysis several phenomena appear. How are we to know which one of them is the phenomenon that we were looking for? And if we know the answer to this question, why then is it necessary to analyse at all?

of reading and that a person is reading if and only if this content, namely the MRP, is present. We shall now show that the process characterized by the presence of the MRP cannot be reading.† First of all, if reading is a particular experience "then it becomes quite unimportant whether or not you read according to some generally recognized alphabetical rule" (165). One is reading if and only if one is experiencing the MRP; nothing else is of any importance. That implies, however, that no distinction can be drawn between reading and believing that one is reading (cf. 202), or, to put it in another way, that anybody who believes that he is reading is entitled to infer that he *is* reading. The important task of a teacher would, therefore, consist in schooling the receptivity of his pupils (232); reading would be something like listening to inner voices in the presence of printed paper and acting in accordance with their advice (233). That different people who are reading the same text agree in the sounds they utter would be miraculous (233). †Our assumption that reading is a mental act leads, therefore, to the substitution of miracles for an everyday affair. It leads also to the substitution for a simple process (uttering sounds in the presence of printed paper) of a more complicated one (listening to inner voices in the presence of printed paper) i.e. it misses the aim of explaining the process of *reading*.†[10]

(e) But does introducing inner voices really solve our problem, namely, to explain why people read correctly and to justify our own reading of a text in a certain way? Usually we simply read off the sound from the letters. Now we want to be justified, and we think that a mental content might justify our procedure. But if we do not trust the signs on the paper, then why should we trust the more ethereal advice of intuition, or of the mental content which is supposed to be the essence of reading? (232, 233).

4. †What conclusions are to be drawn from this analysis? First of all, it appears impossible to discover the essence of a thing in the way that is usually supposed, i.e. *T*(4) seems to be inapplicable. But if that is the case, the correctness of the analysis can no longer be checked in the usual way. There is no criterion for deciding whether a statement like '"*A*" stands for *a*' or 'the sentence "*p*" designates the proposition that *p*' is true or not; and there is no way to decide whether a certain sign is meaningful, either. But usually we are not at all troubled by such questions. We talk and solve

[10] In presenting the idea to be criticized we assumed, as in *T*(4), that the *MRP* is also the reason for our uttering the sounds we utter. The criticism developed in the text applies also to the idea that in calculating we are guided by intuitions (Descartes' theory): It is said that the perception of '2 + 2' is followed by a non-perceptual mental event which advises us how to behave in the sequel; it whispers, as it were, into our mind's ear, 'Say 4!' But the idea cannot explain why we calculate as we do. For instead of explaining the process of obeying a rule (the rule of the multiplication table) it describes the process of obeying a kind of inspiration. In the case of an inspiration, I *await* direction. But I do not await inspiration when saying that 2 + 2 are 4 (232).

(mathematical, physical, economic) problems without being troubled by the fact that there is apparently no possibility of deciding whether or not we are acting reasonably, whether or not we are talking sense. But isn't that rather paradoxical? Isn't it rather paradoxical to assume that an expression which we constantly use to convey, as we think, important information is really without meaning, and that we have no possibility of discovering that fact? And since its being meaningless apparently does not at all affect its usefulness in discourse (e.g. for conveying information), doesn't that show that the presuppositions of the paradox, in particular *T*(1), (2), need reconsideration?[11]

5. †A great deal of the *Philosophical Investigations* is devoted to this task.† The phenomena of language are first studied in primitive kinds of application "in which one can command a clear view of the aim and functioning of words" (5; 130). The primitive, rudimentary languages which are investigated in the course of these studies are called "*language-games*". Let us consider one such language-game: it is meant

> to serve for communication between a builder A and an assistant B. A is building with building-stones: there are blocks, pillars, slabs and beams. B has to pass the stones, and that in the order in which A needs them. For this purpose they use a language consisting of the words 'block', 'pillar', 'slab', 'beam', A calls them out; – B brings the stone which he has learned to bring at such and such a call. – Conceive this as a complete primitive language [2].

Consider first of all how A prepares B for the purpose he is supposed to fulfil. "An important part of the training will consist in the teacher's pointing to the objects, directing the [assistant's] attention to them and at the same time uttering a word, for instance the word 'slab' as he points to that shape" (6; "child" replaced by 'assistant'). This procedure cannot be called an ostensive definition, because the assistant who at the beginning is supposed to be without any knowledge of any language cannot as yet *ask* what the name is (6); which shows that teaching a language can be looked at as "adjusting a mechanism to respond to a certain kind of influence" (497; cf. 5). Finally the assistant is able to play the game, he is able to carry out the orders given to him by the builder A. Let us now imagine that A teaches B more complicated orders – orders which contain colour-names, number-words ('four red slabs!') and even orders which contain what one would be inclined to call descriptions ('Give me the slab lying just in front of you!'), etc.

[11] There is another presupposition as well, namely that in section 3 *all* possibilities of experiencing the essence have been considered. Clearly this assumption cannot be proved. But one thing is certain: we considered all possibilities of experiencing the essence which have so far been treated by philosophers who follow theory *T*. Cf. H. Gomperz, *Weltanschauungslehre* II, 140ff, where medieval realism about concepts is criticized by arguments like Wittgenstein's. Cf. also n.23 below.

> Now, what do the words of the language *signify*? – What is supposed to shew what they signify, if not the kind of use they have? And we have already described that. So we are asking for the expression 'This word signifies *this*' to be made a part of the description. In other words the description ought to take the form 'the word . . . signifies . . .' . . . But assimilating the descriptions of the uses of words in this way cannot make the uses themselves any more like one another. For, as we see, they are absolutely unlike [10].

†Compare, for example, the way in which the word 'four' is used with the way in which the word 'slab' is used within the language-game in question. The difference in the uses of the two words comes out most clearly when we compare the procedures by means of which their respective uses are taught. A child who is to count correctly has first to learn the series of numerals by heart; he has then to learn how to apply this knowledge to the case of counting, for example, the number of apples in a basket. In doing so, he has to say the series of cardinal numbers and for each number he has to take one apple out of the basket (cf. 1). He has to be careful not to count one apple twice or to miss an apple. The numeral which according to this procedure is coordinated with the last apple is called 'the number of apples in the basket'. This is how the use of numerals is taught and how numerals are used in counting. Compare with this the use of a word like 'slab'. It is taught by simple ostension: the word 'slab' is repeatedly uttered in the presence of a slab. Finally the child is able to identify slabs correctly within the language-game it has been taught. Nothing is involved which has any similarity to the counting procedure which was described above. The application of the word itself to a concrete object is much simpler than the application of a number-word to a collection whose cardinal number cannot be seen at a glance. This application does not involve any complicated technique; a person who understands the meaning of 'slab' is able to apply this word quite immediately.†

Let us now imagine that somebody, following *T*(1), should argue in this way: it is quite clear: 'slab' signifies slabs and 'three signifies three . . . every word in a language signifies something (cf. 3). According to Wittgenstein, he has

> so far said *nothing whatever*; unless [he has] explained exactly *what* distinction [he] wish[es] to make. (It might be of course that [he] wanted to distinguish the words of [our] language[-game] from words 'without meaning' [13].

> Imagine someone's saying: '*All* tools serve to modify something. Thus the hammer modifies the position of the nail, the saw the shape of the board and so on.' – And what is modified by the rule, the glue-pot, the nails? – 'Our knowledge of the thing's length, the temperature of the glue and the solidity of the board.' – Would anything be gained by this assimilation of expressions? [14].

6. †Our example and its interpretation suggest an instrumentalist theory of language.[12] The orders which A gives to B are instruments in getting B to act in a certain way. Their meaning depends on how B is supposed to act in the situations in which they are uttered. It seems reasonable to extend this theory – which is a corollary to T', soon to be described – to language-games which contain descriptive sentences as well. The meaning of a descriptive sentence would then consist in its role in certain situations; more generally, within a certain culture (cf. 199, 206, 241, 325, p. 226). Wittgenstein has drawn this consequence, which is another corollary of T':

> What we call '*descriptions*' are instruments for particular use. Think of a machine-drawing [which directs the production of the machine drawn in a certain way], a cross-section, an elevation with measurements, which an engineer has before him. Thinking of a description as a word-picture of facts has something misleading about it: one tends to think only of such pictures as hang on the walls: which seem simply to portray how a thing looks, what it is like. (These pictures are as it were idle) [291].

And quite generally: "Language is an instrument. Its concepts are instruments" (569). This idea has an important consequence. Instruments are described by referring to how they work. There are different kinds of instruments for different purposes. And there is nothing corresponding to the ethereal meanings which, according to $T(1)$, are supposed to make meaningful the use of *all* instruments alike. "Let the use of words teach you their meaning" (p. 220) is to be substituted for $T(4)$; and this now seems to be the new theory, T'. But in order to appreciate the full importance of T' we have first of all to consider the following objections, which seem to be inevitable. In talking, ordering, describing, we certainly use words and get other people to act in a certain way (to revise their plans which we show to be unreasonable, to obey our wishes, to follow a certain route which we point out to them on a map). But the description of the meanings of the elements of a language-game is not exhausted by pointing to the way in which we use those elements and the connection of this use with our actions and other people's. For in uttering the words and the sentences we *mean* something by them, we want to express our thoughts, our wishes, etc. (cf. 501). It is "our *meaning* it that gives sense to the sentence. . . . And 'meaning it' is something in the sphere of the mind" (358; cf. $T(4)$). What we mean

[12] Or an intuitionist (pragmatist, constructivist) theory of language, the expressions 'intuitionist' or 'pragmatist' being used in the way in which they serve to describe one of the present tendencies as regards foundations of mathematics. I am inclined to say – and there is strong evidence in favour of this view – that Wittgenstein's theory of language can be understood as a constructivist theory of meaning, i.e. as constructivism applied not only to the meanings of mathematical expressions but to meanings in general. Cf. H. Poincaré, *Derniers Pensées* (German edn. Leipzig, 1920) 43ff, and especially Paul Lorenzen, 'Konstruktive Begründung der Mathematik' *Math. Z.*, 53 (1950), 162ff. Cf. also *Philosophical Investigations*, p. 220: "Let the *proof* teach you *what* was being proved."

seems to be independent of the way we use our words and the way other people react to our utterances (cf. 205, and again *T*(4)). Moreover, the meanings of our utterances, being hidden beneath the surface of the various ways in which we use their elements, can only be discovered by looking at the mental pictures, the presence of which indicates what we mean by them. A person who wants to understand has, therefore, to grasp this mental picture. "One would like to say: 'Telling brings it about that [somebody else] *knows* that I am in pain [for example]; it produces this mental phenomenon: everything else [in particular whether "he does something further with it as well" – e.g. looks for a physician in order to help me] is inessential to the telling'" (363). "Only in the act of understanding is it meant that we are to do *this*. The *order* – why, that is nothing but sounds, ink-marks" (431). Meaning and understanding are, therefore, mental processes.

†Apparently this idea makes it necessary to give an account of meaning which is independent of the description of the way in which signs are used within a certain language-game. Another great part of the *Philosophical Investigations* is devoted to showing that this is not the case. A careful analysis of the way we use phrases such as 'A intends to . . .', 'A means that . . .', 'A suddenly understands that . . .', shows that in trying to account for this use we are again thrown back on a description of the way we use certain elements of the language-game in which those expressions occur, and the connection of this use with our actions and other people's.

7. †(A) The meaning that we connect with a certain sign is a mental picture. We do not look into the mind of a person in order to find out what he is really saying. We take his utterances *at their face value*, for example, we assume that, when saying 'I hate you' he is in a state of hating. "If I give anyone an order I feel it to be *quite enough* to give him signs. And I should never say: this is only words. Equally, when I have asked someone something and he gives me an answer (i.e. a sign) I am content – that was what I expected – and I don't raise the objection: but that's a mere answer" (503). On our present view, this attitude is easily shown to be superficial. For it might be that on looking into the speaker's soul (or mind) we discover something quite different, for example, love in the person who said 'I hate you'.

†Now two questions arise about this procedure. First, why trust the language of the mind (one wonders what kind of language this may be) when we do not trust the overt language, i.e. the sentence 'I hate you'? (cf. e.g. 74 and all the passages on the interpretation of rules: 197ff). For whatever appears to be found in the mind can be interpreted in various ways, once we have decided *not* to proceed as we usually do, i.e. not to take parts of a certain language-game which we are playing at their face value.

Secondly, let us assume that somebody who really loves a certain person tells her that he hates her.[13] Does this fact make 'I hate you' mean the same as 'I love you'? Or imagine a person, who abounds in slips of the tongue (or is at the moment rather occupied with a difficult problem and so not listening attentively), giving what we consider to be a wrong or an irrelevant answer. Doesn't that reaction of considering his answer as irrelevant show that what he says is thought to be meaningful independently of what he is thinking? For we don't say 'He certainly gave the right answer; what he said was accompanied by the right thought-processes', but rather 'He gave a quite irrelevant answer; maybe he didn't understand our question or expressed himself wrongly.' Or "suppose I said 'abcd' and meant: the weather is fine. For as I uttered the signs I had the experience normally had only by someone who had year-in year-out used 'a' in the sense of 'the', 'b' in the sense of 'weather' and so on. Does 'abcd' now mean: the weather is fine?" (509; cf. 665). How does somebody else find out what I meant by 'abcd'? Of course I can explain to him that 'abcd' means 'the weather is fine'; and I can also indicate how the parts of the first string of signs are related to the parts (the words) of the second string. But it would be a mistake to assume that such an explanation reveals what 'abcd' really means. For from the few words which I intend to be an explanation one cannot yet judge whether an *explanation* has been given or not.

†Of course I *say* '"abcd" means "the weather is fine"' or 'By "abcd" I mean "the weather is fine"', and I have the intention of giving an explanation. But now imagine someone's saying 'Mr A and Mrs B loved – I mean lived – together for a long time.'[14] In this case he does not want to give a definition or an explanation according to which 'love' is supposed to mean the same as 'live'; rather, he committed a slip of the tongue and wanted to correct himself. In certain cases this is clear enough. In other cases it follows, for example, from the fact that 'love' is never again mentioned in connection with Mr A and Mrs B, etc. When, therefore, I say 'By "abcd" I mean "the weather is fine"', it is not yet certain what the case is, whether I intended to give an explanation, or was just awaking from a kind of trance, or whatever else might be the case. The way 'I mean' is to be interpreted follows from the context in which the whole sentence is uttered and from what we find out about the further use of the sign 'abcd' (cf. 686). In order to find out whether 'abcd' really means 'the weather is fine' we have, therefore, to find out how 'abcd' is being used quite independently of any feelings on the part of the person who said 'abcd' and of any explanation

[13] Psychoanalysis has made rather a misleading use of such cases. It has introduced a picture-language (so-called symbols) and interpreted it in such a way that it is not conceivable how the theory could possibly be refuted.

[14] In Freud's *Vorlesungen zur Einführung in die Psychoanalyse* (Berlin, 1933) there are plenty of examples of this kind.

given by him. Of course his explanation may be the starting point of a training in the use of a new language in which 'a', 'b', 'c', 'd' really have the meanings indicated. But note now that 'abcd' makes sense only within this language-game. I cannot mean 'the weather is fine' by 'abcd' before this language-game has been established. I myself could not possibly connect any sense with 'abcd' before the elements of this sign have become meaningful by being made elements of a certain language-game. And even the fact "that I had the experience normally had only by someone who had year-out year-in used 'a' in the sense of 'the', 'b' in the sense of 'weather' and so on" (509) could not make them meaningful; I could not even *describe* this experience as I did just now, because such a description does not yet exist.

†We have to conclude that no mere mental effort of a person A can either make a string of signs mean something different from the meaning it has within a certain language-game of which it is part, played by the people who come into contact with A, or justify its being said that *he* means (intends) something different from everybody else who uses it. This seems rather paradoxical. But let us assume for a moment that two people

> belonging to a tribe unacquainted with games should sit at a chess-board and go through the moves of a game of chess; and with the appropriate mental accompaniments. And if *we* were to see it we should say they were playing chess. But now imagine a game of chess translated according to certain rules into a series of actions which we do not ordinarily associate with a *game* – say into yells and stamping of feet. And now suppose those two people to yell and stamp instead of playing the form of chess that we are used to; and this in such a way that their procedure is translatable by suitable rules into a game of chess. Should we still be inclined to say they were playing a game? [200].

The decision of this question again depends on the situation. Imagine, for example, that their yelling and stamping has an important role within a religious ceremony of the tribe. That any change of procedure is said to offend the gods and is treated accordingly (the offenders are killed). In this case neither the possibility of the translation nor the presence of the chess-feelings in the minds of the participants would turn this procedure into a game of chess (although it is also quite possible to imagine a tribe where people who lose games of chess are thought to be hated by the gods and are killed. But in this case a difference will be made between games and religious procedures by the fact, for example, that only priests are admitted to the latter, or that different expressions are used for describing them, which is missing in our case). On the contrary, the strange mental state of those who are troubled by chess-feelings would be an indication either of insanity (cf. n.5 above) or of lack of religious feeling.

†Now we can turn round our whole argument and look at the people who are sitting at a chess board and moving the pieces. Are they really playing

chess? We see now that the inspection of their minds does not help us: they might be queer people, thinking of chess when they are performing a religious ceremony. Their assertion that they are playing chess, even, is not necessarily helpful, for it might be that they heard the words from somebody else and misinterpreted them to mean sitting in front of the board and making arbitrary moves with the pawns. The fact that they are using a chess board does not help us either, for the board is not essential to the game. What, then, is essential? The fact that they are playing according to certain *rules*, that they follow the rules of the chess game. Applying this result to the meaning of sentences in general we arrive at the idea that "if anyone utters a sentence and *means* or *understands* it he is operating a calculus according to definite rules" (81). Thus in analysing the concepts of meaning, understanding, thinking, etc. we finally arrived at the concept of *following a rule*. But before turning to that concept we have to get more insight into the concepts just mentioned, and especially into the concept of *intention*.[†]

(B) It is the "queer thing about *intention*, about the mental process, that the existence of a custom, of a technique, is not necessary to it. That, for example, it is imaginable that two people should play chess in a world in which otherwise no games existed; and even that they should begin a game of chess – and then be interrupted" (205). The underlying idea is the same, as in the case of meaning; just as we can attach meaning to a sign by just connecting its use with a certain image which we voluntarily produce, we can also intend to do something by producing a certain mental picture. But how, we have to ask, is it possible to find out whether or not A, who just announced his intention of playing chess, was really intending to do so? Surely chess is defined by its rules (cf. 205). Should we therefore conclude that the rules of chess were present in the mind of A when he uttered his intention? (205).

[†]Investigation similar to that of (Ca) of section 3 will show that not every act of intending to play chess is accompanied by a special mental picture which is characteristic of the intention of playing chess. Of course, the intention to play chess is sometimes present quite distinctly (I have not played chess for a long time, I am a keen chess player, and *now* I want to play chess and won't stop looking until I have found a chess board and a suitable partner). But this is only a *special kind* of intending to play chess (cf. (Cb) of section 3); therefore its characteristics cannot be the reason for calling other cases cases of intending to play chess, cases, for example, in which these characteristics are completely absent. But if we assume, on the other hand, that A has a perfect copy of the rules of chess before his inner eye, must he necessarily follow the features of this copy in such a way that the results will be a game of chess? Is it not possible that he either interprets them in an unusual way, or that in going over from the reading of his mental

picture to the outer world (the chess board, his actions in front of the chess board), he automatically makes a kind of translation, so that finally he is not doing what one would be inclined to call 'playing chess' (cf. 73, 74, 86, 139, 237)? And should we still say that he is intending to play chess just because, somewhere in the chain of events which in the end lead to his actions, a copy of the rules of chess enters in? Of course, we could *interpret* this copy as we are used to do. But is *he* interpreting it in the same way? And even if he could tell us how he is interpreting it do we know how to take his explanation? We see that "interpretations *by themselves* do not determine meaning" (198). We have simply to wait. And if he really acts in such a way that he regards playing chess – as we understand it – as a fulfilment of his intention, then we may say that he intended to play chess. But if it turned out that he did not know how to play chess or that, apparently intending to play chess, he sat down at the chess board and made irregular moves, we should under certain circumstances conclude that he had wrong ideas as to his intentions. Of course the phrase 'under certain circumstances' has to be inserted. For it is perfectly possible that A, intending to play chess, was introduced to a person he did not like and, with the intention of avoiding playing chess with him, acted as if he did not know the rules of chess or as if he had never intended to play chess. But what has to be criticized is the idea that such a difference might be found out by inspecting his mind (or soul) and by reading off his intention from his mental processes. It is his further actions (talking included), as well as his personal history, which teach us how we are to take his first utterance – that he intended to play chess. But as it now turns out that our criteria for deciding whether a person, A, intends to play chess or not are "extended in time" (cf. 138), we have to conclude that intending to play chess cannot be a mental event which occurs at a certain time. *Intending is not an experience* (cf. p. 217): it has "no experience-content. For the content (images for instance) which accompany and illustrate [it] are not the . . . intending" (p. 217).

†(C) The same applies to understanding.† Let us examine the following kind of language-game (143ff.): When A gives an order, B has to write down series of signs according to a certain formation-rule. The orders are of the kind '1, 2, 3, . . .!' or '2, 4, 6, 8, . . .!' or '2, 4, 9, 16, . . .!' or '2, 6, 12, 20, 30, 42, . . .!' etc. B is supposed to continue the series in a certain way, i.e. he is supposed to write down the series of numerals in the first case, the series of the even numerals in the second case, etc. First of all, A will teach B the rules of the language-game. He will then give orders to B, in order to check B's abilities. He will finally state that B has mastered the system, that he understands it. It should be clear that, when used in this way, 'understanding' cannot signify a mental phenomenon. For we also say that B understands (is master of) the language-game just explained when lying on his bed and sleeping (cf. 148). But the mental-act philosopher is ready with a

new expression; he speaks of a *subconscious* mental phenomenon, i.e. he says that B, although dreaming perhaps of beautiful women, is nevertheless subconsciously thinking of the new language-game and its rules.

The objections to this idea are obvious. Whether subconscious or not, the alleged thinking-process may or may not determine the actual behaviour of B (cf. (B) of section 7); i.e. B may not be able to carry out the orders of A although a clever psychologist has found out that the thinking-process which is supposed to accompany his ability to obey the orders is present. We shall not say in this case that B has mastered the game, that we have discovered a special case of mastering the game (cf. (Cb2) of section 3) we shall simply say that he had not mastered it although he or the psychologist thought he had. This objection being accepted, it might be said that

> knowing the game is a state of the mind (perhaps of the brain) by means of which we explain the *manifestations* of that knowledge. Such a state is called a disposition. But there are objections to speaking of a state of the mind here, inasmuch as there ought to be two different criteria for such a state: a knowledge of the construction of the apparatus quite apart from what it does [149].

What the apparatus does is in our case the actual behaviour of B when he receives certain orders.

But there is a second way in which the word 'understanding' is used. Understanding in this sense is not meant to be understanding of a game as a whole (understanding the rules of chess, i.e. knowing how to play chess) but understanding the meaning of a particular move within the game, for example, understanding the order 2, 4, 6, . . .! "Let us imagine the following example: A writes series of numbers down, B watches him and tries to find a law for the sequence of numbers. If he succeeds, he exclaims: 'Now I can go on!' – So this capacity, this understanding is something that makes its appearance in a moment" (151), and this suggests that 'understanding', used in this way, might mean a mental event. But wait: *do* we find any mental event which is common to all cases of understanding? Imagine that A gave the order 1, 5, 11, 19, 29, . . .! to B and that, upon A's arriving at 19, B said 'I understand'. What happened to B?

> Various things may have happened; for example, while A was slowly putting one number after the other, B was occupied with trying various algebraic formulae on the numbers which had been written down. After A had written the number 19, B tried the formula $a_n = n^2 + n - 1$; and the next number confirmed his hypothesis. Or again – B does not think of formulae. He watches A writing his numbers down with a certain feeling of tension and all sorts of vague thoughts go through his head. Finally he asks himself: 'What is the series of differences?' He finds the series, 4, 6, 8, 10 and says: Now I can go on. – Or he watches and says 'Yes, I know *that* series' – and continues it, just as he could have done if A had written down the series 1, 3, 5, 7, 9. – Or he says nothing at all and simply continues the series. Perhaps he had what may be called the feeling 'that's easy!' [151].

> We can also imagine the case where nothing at all occurred in B's mind except that he suddenly said 'Now I know how to go on' – perhaps with a feeling of relief [179].

> But are the processes which I have described here *understanding*? [152].

Is it not possible that a person who has the feelings just described is not able to write down the series as it was meant by A? Should we not be inclined to say that he did not really understand? "The application is still a criterion of understanding" (146). It would, therefore, be quite misleading "to call the words ['Now I can go on'] a 'description of a mental state'. – One might rather call them a 'signal'; and we judge whether it was rightly employed by what he [i.e. B] goes on to do" (180).

†Now let us use this example to discuss intention and meaning as well. What if B, in carrying out the order 2, 4, 6, 8, . . .! wrote 1,000, 1,004, 1,008, 1,012, etc. (cf. 185)? Of course A will say: 'Don't you see? You ought to write 2, 4, 6, 8, . . .!' And if that does not lead to a change in the behaviour of B, he will tell him: "What I meant was that [you] should write the next but one number after every number [you] wrote; and from this all those propositions follow in turn" (186). Now several conclusions may be drawn from this situation.† First of all one may be inclined to say that 2, 4, 6, 8, . . .! was an incomplete order and that there was clearly a possibility of misunderstanding (cf. a similar argument in 19). For this order reveals so to speak, only an external character of the series to be written down, namely the character that its first members are '2', '4', '6', etc. And the training of B, too, taught him only an external character of all the series, namely, that they began in a certain way. B has therefore to *guess* how to continue, and of course he may hit upon the wrong guess. But the order 'take the next but one!' seems to be of a different character. It contains so to speak the whole of the series in a nutshell. Understanding *this* order implies knowing the law of development for the whole series. But let us now investigate how the understanding of this order may be taught. Of course A has to write down the series 2, 4, 6, 8, . . . and has to explain to B what 'next but one' means. He does so by comparing this series with 1, 2, 3, 4, . . . and by showing that '4' is the 'next but one to 2', etc. The explanation will therefore be similar to the explanation of 2, 4, 6, 8, . . .! Why, then, should teaching the pupil how to take 'the next but one' remove any possibility of error? On the contrary! We could imagine that B has been taught how to use 2, 4, 6, 8, . . .! but that he does not know what 'the next but one' means. In this case the teacher would have to explain the 'next but one' by referring to 2, 4, 6, 8, . . .! and not the other way round. The same applies to algebraic formulae. Consider a 'difficult' series such as 1, 3, 13, 21, 31, 43, ... It is not easily seen how this series might be continued. If we hear that its algebraic formula is $n^2 - n + 1$ we are able to write down the next numbers at once. But that only

shows that we already know how to apply the algebraic expression, but did not know how to apply 1, 3, 13, 21, 31, 43, . . . if the continuation of this series is ordered. It does not show us an essential quality which, so to speak, contains the whole series in a nutshell. For an onlooker who is unacquainted with the formula as well as with the series will have to learn how to apply the formula in developing series. And the methods of teaching this ability will be similar to the methods of teaching 2, 4, 6, 8, . . .! (cf. 146).

Let us return now to intention. The existence of algebraic formulae for the description of series is misleading in one way: A cannot write down the whole series in order to make himself understood to B. But he can use an algebraic formula or a simple expression, such as 'take the next but one'. He can write down the formula within a few seconds and one is therefore inclined to assume that meaning the series 1, 2, 3, 4, . . . *ad infinitum* can be a mental act which occurs within a few seconds.

> Here I should first of all like to say: Your idea was that that act of meaning the order had in its own way already traversed all those steps; that when you meant it your mind as it were flew ahead and completed all the steps before you physically arrived at this or that one. – Thus you were inclined to use such expressions as: The steps are *really* already taken, even before I take them in writing or orally or in thought [188].[15]

They "are determined by the algebraic formula" (189). But how? Surely thinking of the formula cannot help us (cf. 146), for one and the same formula may be used for different purposes (think of the different use which is made of the formula $a + b = b + a$ in different parts of mathematics: in class theory it means the commutativity of class disjunction; in algebra it is used for expressing the commutativity of algebraic addition; in number theory it is used for expressing a general property of numbers; in lattice theory it has still another meaning and likewise in group theory, etc.). The imagining of the formula (if it ever does occur) must be connected with a certain application of the formula in order to provide us with the knowledge of its meaning and with the knowledge of the speaker's intention in using it. And as it is always possible to apply a formula in many different ways we have to observe how it is applied in a particular case, by a particular mathematician, in order to determine his way of using the formula and thus *what he means* when he utters the formula. But the use of a formula is "extended in time" (138). And therefore, since following up this use is one of the criteria we employ to find out what is meant by A when he writes down a certain formula, we cannot say that meaning something is a mental event. "It may now be said: 'The way the formula is meant determines

[15] Here is the core of Wittgenstein's criticism of the so-called Cantorian (cf. Poincaré, *Derniers Pensées*) interpretation of mathematics. This criticism (it is developed in detail in his mathematical writings; in the *Philosophical Investigations* there are only a few passages, cf. 352) is another corollary of *T*.

which steps are to be taken.' What is the criterion for the way the formula is meant? It is for example, the kind of way we always use it, the way we are taught to use it" (190).[16]

(D) Another criticism of the idea that meaning is a mental activity derives from the fact that sometimes it is calculation that decides the question whether a sentence is meaningful or not. Consider the sentence "I have *n* friends and $n^2 + 2n + 2 = 0$.' Does this sentence make sense?" (513). Assuming that a sentence is made meaningful by connecting its utterance with a certain mental content, we should conclude that there is no difficulty; we have only to look for the mental picture behind it, and that will teach us how to judge. But that is not the case; we are even inclined to say that we do not yet know whether anybody will be able to connect any meaning with the sentence, i.e. according to the theory we are discussing at present, whether anybody is justified in connecting an image with the utterance of this sentence. We have first to find out whether the sentence conforms to certain general rules (the number of friends can neither be negative nor imaginary) and we do so by calculating. We also cannot say at once whether we understand or not; we have first to find out whether there is anything to be understood; i.e. whether we understand or not can again be found out by a process of calculation only. One has, therefore, to realize that "we *calculate*, operate with words and in the course of time turn them sometimes into one picture, sometimes into another" (449).

(E) Result: Meaning, understanding, intending, thinking (and, as we may add, remembering, loving, hoping[17]) are *not mental activities*. The criteria by which we decide whether or not A is thinking of. . ., intending to do . . ., meaning . . ., etc. do not relate only to the moment of the intention, the thought, the understanding. We cannot say "A intended . . . because" and point to a process which accompanies his utterances or his (apparently intentional) behaviour. "For no *process* could have the consequences of [intending]" (cf. p. 218).

8. †The last section was devoted to the discussion of a possible objection against an instrumentalist theory of language, as it seems to be suggested by Wittgenstein (cf. section 7). The objection was founded on the idea that

[16] Cf. also 693: "'When I teach someone the formation of the series . . . I surely mean him to write . . . at the hundredth place.' – Quite right; you mean it. And evidently without necessarily even thinking of it. This shews you how different the grammar of the verb 'to mean' is from that of 'to think'. And nothing is more wrong-headed than calling meaning a mental activity!"

[17] "What is a *deep* feeling? Could someone have the feeling of ardent love or hope for the space of one second – *no matter what* preceded or followed this second? What is happening now has significance – in these surroundings. The surroundings [the history of the event included; cf. the words "what preceded"] give it its importance" (583; cf. 572, 584, 591, 614ff, esp. 638: "If someone says 'For a moment. . .' is he really only describing a momentary process? – But not even the whole story was my evidence for saying 'For a moment . . .'").

words are meaningful because we *mean* something when uttering them, and that quite independently of the way in which those words are used. But it turned out that in deciding whether somebody is really meaning something when uttering a sentence we are thrown back on observation of the way he uses certain elements of speech and that, therefore, an account of meaning can and must be given within the instrumentalist interpretation of language. Meaning is not something that needs consideration *apart* from the description of the way certain expressions are used by the speaker or by other people with whom he is trying to communicate. At the same time a tendency was discovered, namely the tendency "to hypostatize feelings where there are none" (598).[18] No objection to the instrumentalist interpretation seems to be left, but one: when playing a language-game we certainly obey certain *rules*. Thus the idea is suggested "that if anyone utters a sentence and *means* or *understands* it, he is operating a calculus according to definite rules" (81), and the rules seem to be something which *directs* the activities within a language-game, which therefore cannot be described in terms which are useful for describing the working of the language-game itself. It is this idea which we have to treat last. The discussion of this idea in the *Philosophical Investigations* is interwoven with the discussion of the other ideas treated in the book because there are arguments which apply to several ideas at once.[†]

Assuming that in talking, calculating, etc. we are acting in accordance with certain rules leads at once to the following question: "How am I able to obey a rule?" (217). For, on the one hand, it seems to be the case that "the rule, once stamped with a particular meaning, traces the lines along which it is to be followed through the whole space . . . all the steps are already taken" (219). But "if something of that sort really were the case, how would it help?" (219). For is there not always the possibility of interpreting the rule in a different way? And how are we to know which interpretation is the right one? Once the rule is separated from our activity it seems impossible that it can determine this activity any more. For it may try to make itself known to us by mental events ('grasping' the rule), by a book which contains all rules of the language-game to be played, etc. In any one of those cases we can proceed in many different ways depending on how we interpret, i.e. how we use, the mental picture, the book, etc. in the course of our further activities (cf. 73, 74, 86, 139, 237). Thus it seems that "any course of action [can] be determined by a rule because every course of action [can] be made out to accord with the rule" (201; "could" replaced by 'can').

But "What this shows is that there is a way of grasping a rule which is *not* an *interpretation*, but which is exhibited in what we call 'obeying the rule' and

[18] Cf. 295: "When we look into ourselves as we do philosophy, we often get to see just a picture. A full-blown pictorial representation of our grammar. Not facts; but as it were illustrated turns of speech."

'going against it' in actual cases" (201). That will become clear from the following example (cf. 454): "A rule stands there like a sign-post. Does the sign-post leave no doubt open about the way I have to go? Does it show which direction I am to take when I have passed it?" (85). How do I know which direction I have to go? "If that means 'have I reasons?' the answer is: My reasons will soon give out. And then I shall act, without reasons" (211). "When someone whom I am afraid of orders me [to follow the sign-post], I act quickly, with perfect certainty, and the lack of reasons does not trouble me" (212, with "to continue the series" replaced by 'to follow the sign-post'). "When I obey a rule, I do not choose. I obey the rule *blindly*" (219). Let us now assume a land where everybody, on seeing a signpost: →, follows it in this direction: ←; where children are advised to follow the signpost in the way indicated, where foreigners who are in the habit of going → when they see a signpost like this: → are taught that they are acting wrongly, that '→' means 'go ←.' Should we say that the inhabitants of our imaginary country are misinterpreting the signpost? Obviously this would not be the right description of the situation, for without being related to human activities (language-games included) the signpost is a mere piece of matter and the question as to its *meaning* (and therefore the question as to whether a certain interpretation is the right one) does not arise at all.

Now it is using the signpost in a certain way, i.e. behaving in a certain way in the presence of the signpost, that gives a meaning to it and that separates it from the other parts of nature which are meaningless in the sense that they are not parts of human language-games. But *behaving in this way* is also called *obeying the rules*. "And hence also 'obeying a rule' is a practice. And to *think* one is obeying a rule is not to obey a rule. Hence it is not possible to obey a rule 'privately': Otherwise thinking one was obeying a rule would be the same thing as obeying it" (202).

Apply this to language-games in general. It follows that "to obey a rule, to make a report, to give an order, to play a game of chess, are *customs* (uses, institutions)" (199) and "not a hocus-pocus which can be performed only by the soul" (454). "To understand a sentence means to understand a language. To understand a language means to master a technique" (199). And so we are back at the instrumental interpretation of language: "Every sign *by itself* seems dead. *What* gives it life? – In *use* it is *alive*. Is life breathed into it then? – Or is the *use* its life?" (432). And questions of meaning, of understanding, of following a rule are to be treated by taking into account the *use* of signs within a certain language-game.

9. †Thus we arrive at the following result. According to *T*, meanings are objects for which words stand. Rules are of a similar ethereal character. Understanding the meanings, grasping the rules, is an activity of the mind, which is the organ for finding our way about in the realm of meaning as the

senses are organs for finding our way about in the physical world. We found that either there is no representation of the meanings or the rules in the mind or, assuming that a representation does exist, that it cannot determine the way in which we proceed because there are always many possibilities of interpretation. According to *T'* the meaning of the elements of a language-game emerges from their *use*, and that use belongs to a quite different category from a single mental event or a mental process, or any process whatever (cf. 196).

†Now a sign can be part of different language-games just as a button can be used in a game of chess (instead of a pawn, e.g. which has been lost) or a game of draughts. Do we try in this case to abstract from the differences between these two kinds of use in order to discover a common quality which will explain to us how it is possible for the button to function both as a pawn and as a piece in draughts? The question does not arise because it seems obvious that the button *changes* its function according to the game within which it is used. But in the case of a language-game, theory *T* seduces us into thinking that the sign '2', for example, is, in any case of its use within language, connected with a single element, its meaning, and that the varieties of its use ('Give me *two* apples!' as said in a grocery; $\int_0^2 x^3 dx = 4$; '*Two* hours ago I met him in the street'; 'The number of solutions of the equation $x^2 + 5x + 4 = 0$ is *two*') are only a superficial aspect. Once this idea has been dropped, once it has been realized that the meaning of a sign is constituted by its use within a certain language-game, words can be looked at as the button was above. And instead of trying to grasp the *essence* of a thing which is to explain the varieties of the use of the sign which stands for the thing, we ought simply to describe the language-game of which the sign is part. "We must do away with all *explanation*, and description alone must take its place" (109). "Our mistake is to look for an explanation where we ought to look at what happened as a 'proto-phenomenon'. That is where we ought to have said: *This language-game is played*" (654). "Look at the language-game as the *primary* thing. And look on the feelings, etc. as you look on a way of regarding the language-game, as interpretation" (656).[19]

†Wittgenstein's position has not yet been described correctly. Wittgenstein was said to hold a theory, *T'*, which emphasizes the instrumental aspect of language and which points to use in a language-game as the essential thing. And describing the language-game, so one is inclined to say, according to the presentation which has been given so far, is the task of philosophy. From that description quite a few philosophical problems will become clear which seemed hopelessly muddled when seen from the point

[19] Note that the idea of an ideal language becomes obsolete as soon as it has been recognized that all language-games are on a par. Vague concepts, e.g. (cf. 71) cannot be regarded as inadmissible any longer. They have a definite function, and that is all we can demand from them.

of view of theory *T*. Philosophy, then, seems to be the theory of language-games (a kind of general syntax or semantics in Carnap's sense) and *T'* seems to be its most important part. But according to Wittgenstein this assumption would involve a misunderstanding. For the supposed theory of language-games could do no more than enable people to run through the single moves of a game, as a player who is acquainted with the game runs through its moves. But for such a player there is no problem. If he asks, for example, "'How do sentences manage to represent?' – the answer must be: 'Don't you know? You certainly see it, when you use them.' For nothing is concealed" (435). Everything "lies open to view" (92; 126). "Philosophy" therefore "may in no way interfere with the actual use of language; it can in the end only describe it . . . it leaves everything as it is" (124).

†Let us assume that somebody begins to construct a theory of language-games. This theory, if formulated in the terms of *T'*, will be thought to serve as an explanation of how meaning is conferred upon single signs by the way in which these signs are incorporated into a language-game. The theory (or description, as it may also be called) will involve a new kind of use of terms such as 'sentence', 'fact', 'meaning'. But has a useful explanation or description really been found? We must realize that the supposed theory introduces a *new* use of 'meaning', 'fact', 'sentence', etc. If this use involves even a slight deviation from the use of these words within the language-games to be described (explained) the supposed description in fact involves a change in the phenomenon to be described. But if on the other hand the change is a considerable one (and that is to be expected if one is trying to develop a fully fledged instrumentalist philosophy of meaning) a new language-game for the expression 'sentence', 'meaning', etc. has been established and the task of describing the given language-game is not fulfilled either. Thus "we must do away with all *explanation*" and with *T'* as well. The description, however, which Wittgenstein invites us to give instead of the explanation, consists only in "putting the things before us" (126), and as "everything lies open to view, there is nothing to explain" (126). We might therefore say, rather hyperbolically, that the "language disguises thoughts" of the *Tractatus* (4.002) is now replaced by "language is already thought, nothing is concealed".

†But the situation is not quite as simple as that. For there *are* philosophical systems, philosophical theories; and it needs to be explained how it is that they come into existence if "nothing is concealed".

†In describing how philosophical theories come into being, Wittgenstein refers to the fact that "we *do not command a clear view* of the use of our words" (122). Given the answer that nothing is concealed, "one would like to retort: 'Yes, but it all goes by so quick, and I should like to see it as it were laid open to view" (435). On the other hand, "we remain unconscious of the prodigious diversity of all the everyday language-games because the

clothing of our language makes everything alike" (p. 224). "What confuses us is the uniform appearance of words when we hear them spoken or meet them in script and print. For their *application* is not presented to us so clearly" (11). Take the following example: The sentences 'Washington is a city' and 'Two is an even number' are of a similar structure. This suggests that just as in the first case 'Washington' is the name of a real thing, 'two' is the name of a more abstract object, notwithstanding the fact that the uses of the two signs are "absolutely unlike" (10).

> In the use of words one might distinguish 'surface-grammar' from 'depth-grammar'. What immediately impresses itself upon us about the use of a word is the way it is used in the construction of the sentence, the part of its use – one might say – that can be taken in by the ear. And now compare the depth-grammar, say of the word 'to mean', with what the surface-grammar would lead us to suspect. No wonder that we find it difficult to know our way about [661].

This difficulty is the reason why we resort to philosophical theories; why we invent theories of meaning; and why we try to conceive an ideal form behind the complexities of our language-games.

†But it is clear "that every sentence in our language 'is in order, as it is'. That is to say, we are not *striving after* an ideal, as if our ordinary vague sentences had not yet got a quite unexceptionable sense . . . there must be perfect order even in the vaguest sentence" (98). It should also be clear that the "philosophy of logic speaks of sentences and words in exactly the same sense in which we speak of them in ordinary life, when we say, e.g., 'Here is a Chinese sentence' or 'No, that only looks like writing; it is actually just an ornament' and so on" (108). Thus the proper task of philosophy will be to unmask philosophical theories, to "bring words back from their metaphysical to their everyday use" (116), to destroy the "houses of cards" and to clear up "the ground of language on which they stand" (113). And philosophy becomes a "battle against the bewitchment of our intelligence by means of language" (109). This battle is carried through by "assembling reminders for a particular purpose" (127), for the purpose of "seeing connexions" (122); and "different therapies" (133), not "*a* philosophical method" (133), are used in order to finish it victoriously.

†But in these therapies the statement of *T* (or rather of the several corollaries of *T* which have been mentioned so far) plays the most important part. So far we have interpreted the statement of *T* as the exposition of a new (instrumentalist, nominalist, or whatever you like to call it) *theory of meaning*. This interpretation is not unreasonable in itself and taken as such it is a very interesting contribution to traditional philosophy (actually I think that everything that is interesting in the book attaches to the treatment of *T* in this way). But this interpretation would go against the way in which his book is meant to be used by Wittgenstein. That may be seen from the

following considerations: In section 3 the idea was criticized that reading is a mental process. If we stick to *T'* and interpret it as a theory we cannot understand why the discussion in section 3 should be a *criticism*. For we could argue in the following way: Wittgenstein says that the meaning of a word becomes clear from the way in which it is used within a specific language-game. Let us, therefore, look at the language-game which contains both of the expressions 'reading' and 'mental process', and in which the sentence occurs 'Reading is a mental process.' Wittgenstein's presentation – so one would be inclined to say – is a description of certain features of this language-game and includes, of course, the remark that 'mental process' as used *in this language-game* has nothing whatever to do with toothaches.

†But that is not the right account of what Wittgenstein does. Wittgenstein does criticize; but his criticism is of a particular kind. It is not the kind of criticism which is directed, for example, against a wrong mathematical calculation. In the latter case the result of the criticism is that a certain sentence is replaced by its negation or by a different sentence. But Wittgenstein does not want his reader to discover that reading is *not* a mental process. For if 'mental process' is used in a metaphysical way in 'reading is a mental process', it is used just as metaphysically in "reading is not a mental process" (cf. 116). For him "the results of philosophy are the uncovering of one or another piece of plain nonsense and of bumps that the understanding has got by running its head against the limits of language" (119), and his aim is "to teach you to pass from a piece of disguised nonsense to something that is patent nonsense" (464) and in this way to clear up "the ground of language" (119). But that can only mean that "the philosophical problems should *completely* disappear" (133); for if the aim has been reached, "everything lies open to view and there is nothing to explain" (126). This implies that the formulation of *T'* as used within the critical procedure cannot be interpreted as a new theory of meaning, for it is applied with the intention of making the language-games (e.g. that with 'reading') "lie open to view", i.e. lead to a situation where language-games are simply played, without any question arising as to how it is that words become meaningful as part of a certain language-game, etc. That being so, the formulation of *T'* loses its function as soon as "*complete clarity*" has been arrived at. But without a function the signs which are part of the formulation of *T'* are without meaning. Thus one could say of the sentences which are part of *T'*: These sentences "are elucidatory in this way: he who understands me finally recognizes them as senseless . . . (He must so to speak throw away the ladder, after he has climbed up on it.) He must surmount these [sentences] . . .; then he sees the world rightly" (*Tractatus*, 6.54). And seeing the world rightly means playing the language games without being troubled by philosophical *questions* or by philosophical *problems*.†

10. •Note, now, that in the preceding section the idea of the essence has been reintroduced. In traditional philosophy the essence was hidden beneath the various ways of describing it. Now it is the "everyday use" (116) that "has to be accepted", "is given" (p. 226); but this everyday use is likewise hidden, beneath the "houses of cards" of philosophical theories (118),[20] and it too has to be brought to light. Just so, traditional philosophers (i.e. the adherents of theory *T*) tried to bring to light the clear and sharp meanings which were hidden beneath the "muddied" use of the words which stand for them (426). If we assume, now, that in removing those philosophical coverings we finally arrive at "*complete* clarity" (133), we assume that there is a *sharp line* between the "houses of cards" on the one hand and the language-games on which they are built on the other. Now while Wittgenstein usually criticizes the idea that, for example, "there *must* be something common [to games], or they would not be called 'games'" (66; cf (A) of section 3), and points to the fact that if we "look and see" (66) we find a "complicated network of similarities overlapping and criss-crossing" (66), he seems to assume, none the less, that at least philosophical difficulties have something in common, that there is a definite boundary between the card-houses of philosophy and the solid ground of everyday language, such that it becomes possible to "bring words back from their metaphysical use to their everyday use" (116).

•To Wittgenstein we can apply the comment (which he used to characterize the adherents of *T*) that "A *picture* held [him] captive" (115). For if it is the use, the practice, which constitutes meaning, if "what has to be accepted, the given, is . . . *forms of life*" (p. 226), then one may ask why Wittgenstein tries to eliminate theory *T*, which certainly must be regarded as a form of life if we look at the way in which it is used by its adherents. Nevertheless Wittgenstein tries to eliminate this theory as well as other philosophical theories. But this attempt can only be justified by assuming that there is a difference between using a sign (playing a language-game) and proceeding according to theory *T*. The procedures which are connected with theory *T* are supposed not to be taken as parts of a language-game, they constitute a sham-game which is to be destroyed. How is this attitude to be understood?

•I think we can understand it by looking at the ideas which Wittgenstein has about philosophy (at his "*picture*" of philosophy as one might call it, using his own word). This picture is the picture of the *Tractatus*: "The word 'philosophy' must mean something which stands above or below, not beside the natural sciences" (*Tractatus*, 4.111). In the *Investigations* we may replace "natural sciences" by "language-games", and we arrive at: "Philosophy must be something which stands above or below, not beside the language-

[20] "Language disguises the thought" is the position of the *Tractatus* (4.002). One could say that according to the *Investigations*, the (philosophical) thought disguises language.

games"; philosophy *cannot* be a language-game itself; for example, it cannot be theory *T'*. I submit that this idea is still present in the *Investigations* and that it makes it clear why Wittgenstein, having found that a sign can only be meaningful if it is incorporated into a language-game, cannot admit that there are philosophical theories.[21] This observation (as well as others which have not been mentioned[22]) suggests that the *Investigations* (apart from their substitution of language-games for the one language of the *Tractatus*) are after all not as different from the *Tractatus* as they seem to be at first sight. I am even inclined to say (without being able to substantiate this contention at the moment) that the *Investigations* basically contain an application of the main ideas of the *Tractatus* to several concrete problems, the only difference being the use of language-games instead of the language of the natural sciences which formed the theoretical background of the *Tractatus*.

Trying to evaluate the book, we might say that the criticisms of *T* and the statement of *T'* which it contains, as well as the application of this theory to the discussion of concrete problems (remembering, obeying an order, the problem of sensation, etc.), are a great achievement, which, however, has its predecessors.[23] *Here we are within traditional philosophy*. But Wittgenstein wants us to see his criticisms in a different light. In the end we should forget them as well as *T*, we should forget philosophy entirely. Although the formulation of what can be regarded as a *theory* (theory *T'*) led us to the proper understanding of our difficulties, it must not be taken as the formulation of a *theory* but only as a proper means of getting rid of our philosophical troubles. *T'* has, therefore, to disappear together with those troubles. This new idea, which is Wittgenstein's own and which can be found in the *Tractatus* as well, is due, first, to the *picture* that philosophy must be something quite extraordinary and, second, to certain difficulties, already mentioned, which could be solved by taking into account the difference between object-language and meta-language (used by Tarski to get rid of similar difficulties, but never recognized by Wittgenstein (cf. 121)). Using this device we find that the philosophical language-games do not necessarily disturb the language-games they are supposed to describe.

[21] There are some passages which seem to contradict this interpretation of Wittgenstein's views, e.g. "If one tried to advance *theses* in philosophy it would never be possible to question them, because everyone would agree to them" (128), according to which philosophical theses are not meaningless, but *trivial*.
Added 1980. Wittgenstein's main point, viz. that principles, general ideas, metaphysical dreams receive content through a practice, may be retained provided we permit the researcher to build up new practices and do not demand that he restrict himself to existing practices entirely. Cf. also *A M*, 31f.

[22] Cf. the similarity of "shows itself" in the *Tractatus* and "lies open to view" in the *Investigations*.

[23] Cf., e.g. H. Gomperz, *Weltanschauungslehre*, II, where further references are given; E. Mach, *Erkenntnis und Irrtum*, 3rd edition (Leipzig, 1917), 126ff; D'Alembert, *Traité de dynamique* (1743); the tenets of the various nominalistic schools, old and new, and especially the work of Brouwer and of his school.

We also find that philosophy is not necessarily on a level with the language-games it is about. On the contrary, the assumption that the philosophical language-games are on a level with the language-games they deal with leads to contradictions. This solution would not agree with Wittgenstein's, but it would retain several elements of his philosophy: (1) his criticisms of T; (2) his statement of T'; (3) his observation that language-games may be disturbed by other language-games which are supposed to explain or to describe them. It would, however, interpret the statement of T' as a special theory of meaning and formulate it by taking account of the difference between object-language and meta-language. It would be possible still to have philosophical theories and philosophical problems without being open to Wittgenstein's criticisms, except perhaps the one criticism that the distinction introduced is purely artificial.*

Added 1980

Wittgenstein's arguments can be applied to important issues of rationalism and scientific method. Scientific rationalists assume that scientific practice, and the practice of thinking in general, is based on simple laws and standards and *must be* based on such standards, and that both can be treated exhaustively by discussing the simple slogans used for expressing them and the logical relations between the slogans: one can understand science, and for that matter any fruitful line of thought, without participating in it. Here Wittgenstein would object – and, to my mind, with complete justification – that slogans receive their meaning only in connection with the practice of science, just as the formula

$$\frac{-a \pm \sqrt{(b^2 - 4ac)}}{2a}$$

receives its meaning only in connection with the practice of solving equations. Otherwise they are irrelevant marginal remarks. (The actual procedure used in the philosophy of science is as follows: its practitioners construct a minipractice which has a 'logic' of its own but which is as different from scientific practice as a toy automobile differs from a real automobile: the working of real automobiles cannot be explained in this manner.)

One should also notice the close connection between Wittgenstein's 'language-games' or 'forms of life' and Kuhn's 'paradigms'. Neither can be understood on the basis of simple and abstract descriptions. The lack of clarity which philosophers who were brought up on such descriptions have noticed in Kuhn can be explained by the fact that Kuhn moves away from traditional conceptions of clarity and precision in order to arrive at a more realistic picture of scientific change: just like a language-game a paradigm is not a well-defined entity but a word for a practice whose elements become known only to those who participate in it.

The similarity should also be noticed between the point of view that Wittgenstein attacks (and which I describe in the first section of ch. 1) and the new abstract traditions described in sections 1ff of ch. 1.

That understanding a practice is impossible without participation was noticed by some students of the social sciences but was described by them in the mentalistic terminology of their times ('Einfühlen'; 'Verstehen'; and so on). They were correct in assuming that in order to understand a practice one must come closer to it than one does when using general hypotheses (cf. Wittgenstein's comments on the role of formulae in mathematics and his criticism of the idea that such formulae give a condensed and comprehensible account of a temporally extended practice; in these remarks he refutes all attempts at a 'theoretical' account of the social sciences, Popper's poor *Poverty of Historicism* (London, 1968) included). They erred in assuming that a coming closer involves mental processes and receives content only through them. They further erred when assuming that a coming closer of the kind described is important only for the social sciences: it is essential for all sciences, physics and mathematics included. Paradoxically speaking one may say that for Wittgenstein (and Brouwer, who quite explicitly spoke in this manner) even mathematics is a social science, and that *there are no natural sciences* in the customary sense.

Finally, a critical comment on Wittgenstein's idea of philosophy. Wittgenstein assumes that philosophers want to provide a theory of already existing things, and he is correct in pointing out that what exists is much more complicated than any philosophical theory. However, philosophical theories have not merely reflected things but have changed them, i.e. the (sham) conflict between theory and practice was resolved by a change of practice. This fact refutes the idea that philosophers, and for that matter all mythmakers, only erect castles in the air, and introduces a fruitful relativism of the kind explained in my *Erkenntnis für Freie Menschen* (Frankfurt, 1980).

8

Consolations for the specialist*

'I have been hanging people for years, but I have never had all this fuss before.' (Remark made by Edward 'Lofty' Milton, Rhodesia's part time executioner on the occasion of demonstrations against the death penalty.) 'He was' – says *Time* magazine (15 March 1968) – 'professionally incapable of understanding the commotion'.

1. INTRODUCTION

In the years 1960 and 1961 when Kuhn was a member of the philosophy department at the University of California in Berkeley I had the good fortune of being able to discuss with him various aspects of science. I have profited enormously from these discussions and I have looked at science in a new way ever since.[1] Yet while I thought I recognized Kuhn's *problems*; and while I tried to account for certain *aspects* of science to which he had drawn attention (the omnipresence of anomalies is one example); I was quite unable to agree with the *theory of science* which he himself proposed; and I was even less prepared to accept the general *ideology* which I thought formed the background of his thinking. This ideology, so it seemed to me, could only give comfort to the most narrowminded and the most conceited kind of specialism. It would tend to inhibit the advancement of knowledge. And it is bound to increase the anti-humanitarian tendencies which are such a disquieting feature of much of post-Newtonian science (cf. ch. 2). On all these points my discussions with Kuhn remained inconclusive. More than once he interrupted a lengthy sermon of mine, pointing out that I had misunderstood him, or that our views were closer than I had made them appear. Now, looking back at our debates[2] as well as at the papers which Kuhn has published since his departure from Berkeley, I am not so sure that this was the case. And I am fortified in my belief by the fact that almost every reader of Kuhn's *Structure of Scientific Revolutions* (Princeton, 1962) interprets him as I do, and that certain tendencies in modern sociology and

* An earlier version of this paper was read in Professor Popper's seminar at the London School of Economics (March 1967). I would like to thank Professor Popper for this opportunity as well as for his own detailed criticism. I am also grateful to Messrs Howson and Worrall for their valuable editorial and stylistic help.

[1] The criticism of some features of contemporary methodology which appears in my *A M* is but one belated after-effect.

[2] Some of which were carried out in the now defunct *Café Old Europe* on Telegraph Avenue and greatly amused the other customers by their friendly vehemence.

modern psychology are the result of exactly this kind of interpretation. I hope that Kuhn will forgive me when therefore I once more raise the old issues and that he will not take it amiss when in my effort to be brief I do this in a somewhat blunt fashion.

2. AMBIGUITY OF PRESENTATION

Whenever I read Kuhn, I am troubled by the following question: are we here presented with *methodological prescriptions* which tell the scientist how to proceed; or are we given a *description*, void of any evaluative element, of those activities which are generally called 'scientific'? Kuhn's writings, it seems to me, do not lead to a straightforward answer. They are ambiguous in the sense that they are compatible with, and lend support to, both interpretations. Now this ambiguity (whose stylistic expression and mental impact has much in common with similar ambiguities in Hegel and in Wittgenstein) is not at all a side issue. It has had quite a definite effect on Kuhn's readers and has made them look at, and deal with, their subject in a manner not altogether advantageous. More than one social scientist has pointed out to me that now at last he had learned how to turn his field into a 'science', by which of course he meant that he had learned how to *improve* it. The recipe, according to these people, is to restrict criticism, to reduce the number of comprehensive theories to one, and to create a normal science that has this one theory as its paradigm.[3] Students must be prevented from speculating along different lines and the more restless colleagues must be made to conform and 'to do serious work'. *But is this what Kuhn wants to achieve?*[4] Is it his intention to provide a historico-scientific justification for the ever-growing need to identify with some group? Does he want every subject to imitate the monolithic character of, say, the quantum theory of

[3] See, e.g. Reagan 'Basic and Applied Research: A Meaningful Distinction?' *Science*, 155 (1967), 1385: he states 'We [that is, we social scientists] are in what Kuhn might call a "pre-paradigm" stage of development in which consensus has yet to emerge on basic concepts and theoretical assumptions.'

[4] Neurophysiology, physiology, and certain parts of psychology are far ahead of contemporary physics in that they manage to make the discussion of fundamentals an essential part of even the most specific piece of research. Concepts are never completely stabilized but are left open and are elucidated now by the one, now by the other theory. There is no indication that progress is hampered by the more 'philosophical' attitude which, according to Kuhn, underlies such a procedure (see 'Logic of Discovery or Psychology of Research' in *Criticism and the Growth of Knowledge*, ed. I. Lakatos & A. Musgrave (Cambridge, 1970), 6). Thus the lack of clarity about the idea of perception has led to many interesting empirical investigations, some of them yielding quite unexpected and highly important results. Cf. S. Epstein, *Varieties of Perceptual Learning* (New York, 1967), esp. 6–18.) On the contrary: we find more awareness of the limits of our knowledge, of its connection with human nature; we find also a greater familiarity with the history of the subject and the ability not only to *record*, but to *actively use* past ideas for the advancement of contemporary problems. Must we not admit that all this contrasts most favourably with the humourless dedication and the constipated style of a 'normal' science?

1930? Does he think that a discipline that has been constructed in this manner is in some ways better off? That it will lead to better, to more numerous, to more interesting results? Or are his followers among sociologists an unintended side effect of a work whose sole purpose is to report '*wie es wirklich gewesen*' without implying that the reported features are worthy of imitation? And if this is the sole purpose of the work, then why the constant misunderstanding, and why the ambiguous and occasionally highly moralizing style?

I venture to guess that the ambiguity is *intended* and that Kuhn wants to fully exploit its propagandistic potentialities. He wants on the one hand to give solid, objective, historical support to value judgements which he, just as many other people, seem to regard as arbitrary and subjective. On the other hand he wants to leave himself a safe second line of defence: those who dislike the implied derivation of values from facts can always be told that no such derivation is made and that the presentation is purely descriptive. My first set of questions, therefore, is: why the ambiguity? How is it to be interpreted? What is Kuhn's attitude towards the kind of following I have described? Have they misread him? Or are they legitimate followers of a new vision of science?

3. PUZZLE SOLVING AS A CRITERION OF SCIENCE

Let us now disregard the problem of presentation and let us assume that Kuhn's aim is indeed to give but a *description* of certain influential historical events and institutions.

According to this interpretation it is the existence of a puzzle-solving tradition that *de facto* sets the sciences apart from other activities. It sets them apart in a 'far surer and more direct' way, in a manner that is 'at once . . . less equivocal and . . . more fundamental',[5] than do other and more recondite properties which they may also possess. But if the existence of a puzzle-solving tradition is so essential, if it is the occurrence of this property that unifies and characterizes a specific and easily recognizable discipline; then I do not see how we shall be able to exclude say, Oxford philosophy, or, to take an even more extreme example, *organized crime* from our considerations.

For organized crime, so it would seem, is certainly puzzle solving *par excellence*. Every statement which Kuhn makes about normal science remains true when we replace 'normal science' by 'organized crime'; and every statement he has written about the 'individual scientist' applies with equal force to, say, the individual safebreaker.

Organized crime certainly keeps foundational research to a minimum[6]

[5] See Kuhn's paper in *Criticism and the Growth of Knowledge*, 7.

[6] Cf. T. S. Kuhn, 'The Function of Dogma in Scientific Research' in *Scientific Change*, ed. A. Crombie (London, 1963), 357.

although there are outstanding individuals, such as Dillinger, who introduce new and revolutionary ideas.[7] Knowing the rough outlines of the phenomena to be expected the professional safebreaker 'largely ceases to be an explorer . . . or at least an explorer of the unknown [after all, he is supposed to know all the existing types of safe]. Instead, he struggles to . . . concretize the known [i.e. to discover the idiosyncrasies of the particular safe he is dealing with], designing much special-purpose apparatus and many special-purpose adaptations of theory for that task'.[8] According to Kuhn failure of achievement most certainly reflects 'on the competence of the [safebreaker] in the eyes of his professional compeers[9] so that 'it is the individual [safebreaker] rather than current theory [of electromagnetism, for example] which is tested':[10] 'only the practitioner is blamed, not his tools';[11] and so we can continue step for step, down to the very last item on Kuhn's list. The situation is not improved by pointing to the existence of *revolutions*. First of all, because we are dealing with the thesis that it is normal science which is characterized by the activity of puzzle solving. And secondly because there is no reason to believe that organized crime will fall behind in the mastery of major difficulties. Besides, if it is the *pressure* derived from the ever-increasing number of anomalies that leads, first to a crisis, and then to a revolution, then the greater the pressure, the sooner the crisis must occur. Now the pressure exerted upon the members of a gang and their 'professional compeers' certainly can be expected to exceed the pressures upon a scientist; the latter hardly ever has to deal with the police. Wherever we look, the distinction we want to draw does not exist.

This of course is no surprise. For Kuhn, as we interpret him now and as he himself very often wants to be interpreted, has failed to do one important thing. He has failed to discuss the *aim* of science. Every crook knows that apart from succeeding at his trade and being popular with his fellow crooks he wants one thing: money. He also knows that his normal criminal activity is going to give him just this. He knows that he will receive the more money and rise the faster on the professional ladder the better he is as a puzzle solver and the better he fits into the criminal community. Money is his aim. What is the aim of the scientist? And, given this aim, is normal science going to lead up to it? Or are perhaps scientists (and Oxford philosophers) less rational than crooks in that they 'are doing what they are doing' without regard to an aim?[12] These are the questions which arise if one wants to restrict onself to the purely descriptive aspect of Kuhn's account.

[7] Dillinger considerably advanced the technique of the bank-holdup by staging dress rehearsals in life size models of the target-banks which he built at his farm. He thereby refuted Andrew Carnegie's 'Pioneering don't pay'.

[8] Kuhn, in *Scientific Change*, 363.

[9] Kuhn in *Criticism and the Growth of Knowledge*, 9; also cf. p. 7 and n.1, p. 5.

[10] *Ibid.*, 5.

[11] *Ibid.*, 7; also cf. Kuhn, *Structure of Scientific Revolutions*, 79.

[12] 'I am doing what I am doing' was a favourite remark of Austin's.

4. THE FUNCTION OF NORMAL SCIENCE

In order to answer these questions we must now consider not only the actual structure of Kuhnian normal science, but also its function. Normal science, he says, is a *necessary presupposition of revolutions*.

According to this part of the argument the pedestrian activity associated with 'mature' science has far-reaching effects both upon the content of our ideas, and upon their substantiality. This activity, this concern with 'tiny puzzles' leads to a close fit between theory and reality, and it also precipitates progress. It does so for various reasons. First of all the accepted paradigm gives the scientist a guide: 'As a glance at any Baconian natural history or a survey of the pre-paradigm development of any science will show, nature is vastly too complex to be explored even approximately at random'.[13] This point is not new. The attempt to create knowledge needs guidance; it cannot start from nothing. More specifically, it needs a theory, a point of view that allows the researcher to separate the relevant from the irrelevant, and that tells him in what areas research will be most profitable.

To this common idea Kuhn adds a specific twist of his own. He defends not only the *use* of theoretical assumptions, but the *exclusive choice* of one particular set of ideas, the monomaniac concern with only one single point of view. He defends such a procedure first, because it plays a role in actual science as he sees it. This is the description-recommendation ambiguity already dealt with. But he defends it also for a second reason that is somewhat more recondite as the preferences behind it are not made explicit. He defends it because he believes that its adoption will in the end lead to the overthrow of the very same paradigm to which the scientists have restricted themselves in the first place. If even the most concerted effort to fit nature into its categories fails; if the very definite expectations created by these categories are disappointed again and again; then we are *forced* to look for something new. And we are forced to do this not just by an abstract discussion of possibilities which does not touch reality, but is rather guided by our own likes and dislikes;[14] we are forced to do it by procedures which have established a close contact with nature, and therefore, in the last resort, by nature itself. The debates of *pre-science* with their universal criticism and their uninhibited proliferation of ideas are 'often directed as much to the members of other schools as . . . to nature'.[15] *Mature science*, especially in the quiet periods immediately before the storm, seems to

[13] Kuhn, in *Scientific Change*, 363.

[14] 'If any one offers conjectures about the truth of things from the mere possibility of hypothesis, then I do not see how any certainty can be determined in any science; for it is always possible to contrive hypotheses, one after another, which are found to lead to new difficulties' (Newton's letter to Pardies, 10 June 1672, in *The Correspondence of Isaac Newton*, ed. H. W. Turnbull (Cambridge, 1959), I, 163–71).

[15] Kuhn, *Structure of Scientific Revolutions*, 13.

address nature itself only and may therefore expect a definite *and objective* answer. In order to get such an answer we need more than a collection of facts assembled at random. But we need also more than an everlasting discussion of different ideologies. What is needed is the acceptance of *one* theory and the relentless attempt to fit nature into its pattern. This, I think, is the main reason why the rejection, by a mature science, of the uninhibited battle between alternatives would be defended by Kuhn not only as a *historical fact*, but also as a *reasonable move*. Is this defence acceptable?

5. THREE DIFFICULTIES OF FUNCTIONAL ARGUMENT

Kuhn's defence is acceptable *provided* revolutions are desirable and provided the particular way in which normal science leads to revolutions is desirable also.

Now I do not see how the desirability of revolutions can be established by Kuhn. Revolutions bring about a *change* of paradigm. But if we follow Kuhn's account of this change, or 'gestalt-switch' as he calls it, it is impossible to say that they have led to something *better*. It is impossible to say this because pre-revolutionary and post-revolutionary paradigms are frequently incommensurable.[16] This I would regard as the first difficulty of the functional argument if used in connection with the remainder of Kuhn's philosophy.

Secondly we have to examine what Lakatos has called the 'fine-structure' of the transition: normal science/revolution. This fine-structure may reveal elements we do not want to condone. Such elements would force us to consider different ways of bringing about a revolution. Thus it is quite imaginable that scientists abandon a paradigm out of frustration and not because they have arguments against it. (Killing the representatives of the status quo would be another way of breaking up a paradigm.[17]) How do scientists *actually* proceed? And how would we *want* them to proceed? An examination of these questions leads to a second difficulty for the functional argument.

In order to exhibit this difficulty as clearly as possible let us first consider the following *methodological problems*. Is it possible to give reasons for proceeding as Kuhn says normal science proceeds, i.e. for trying to stick to a theory despite the existence of *prima facie* refuting evidence, of logical, and of mathematical counter-arguments? And if we assume it is possible to give such reasons, is it then possible to abandon the theory without violating them?

[16] Cf. section 9 below.

[17] This is how religious doctrines or political doctrines were frequently replaced. The principle remains even today, though murder is no longer the accepted method. The reader should also consider Max Planck's remark that old theories disappear because their defenders die out.

In what follows I shall call the advice to select from a number of theories the one that promises to lead to the most fruitful results, and to stick to this one theory even if the actual difficulties it encounters are considerable, the *principle of tenacity*.[18] The problem then is how this principle can be defended, and how we can change our allegiance to paradigms in a manner that is either consistent with it, or perhaps even dictated by it. Remember that we are here dealing with a methodological problem and not with the question of how science actually proceeds. We are dealing with it because we hope that its discussion will sharpen our historical perception and will lead us to interesting historical discoveries.

Now the solution of the problem is quite straightforward. The principle of tenacity is reasonable because theories are capable of development, because they can be improved, and because they may eventually be able to accommodate the very same difficulties which in their original form they were quite incapable of explaining. Besides, it is not at all prudent to put too much trust in experimental results. Indeed, it would be a complete surprise and even a cause for suspicion, if all the available evidence should turn out to support a single theory, even if this theory should happen to be true. Different experimenters are liable to commit different errors and it usually needs considerable time before all experiments are brought to a common denominator.[19] To these arguments in favour of tenacity Professor Kuhn

[18] This formulation of the principle was suggested by an objection which Isaac Levi raised against an earlier version.

The principle of tenacity is formulated in the text should not be confused with Putnam's *rule of tenacity* ('"Degree of Confirmation" and Inductive Logic' in *The Philosophy of Rudolf Carnap*, ed. P. A. Schilpp (Evanston, 1963), 772). For while Putnam's rule demands that a theory should be retained '*unless* it becomes inconsistent with the data' (his italics) tenacity as understood by Kuhn and by myself demands that it should be retained *even if there are data which are inconsistent with it*. This stronger version creates problems which do not appear in Putnam's methodology and which, I suggest, can be solved only if one is prepared to use a multiplicity of mutually inconsistent theories *at any time of the development of our knowledge*. It seems to me that neither Kuhn nor Putnam is prepared to take this step. But while Kuhn sees the need for the use of alternatives (see below) Putnam demands that their numbers be always reduced either to one or to zero (*ibid.*, 770ff).

Lakatos differs from the account given in the text above in two respects. He distinguishes between *theories* and *research programmes*. And he applies tenacity to research programmes only.

Now while I admit that the distinction and the use he makes of it may increase clarity, I am still inclined to stick to my own and much more vague term 'theory' (for a partial explanation of this term, cf. vol. 1, ch. 5, n.5) which covers both Lakatos' 'theories' and 'research programmes', to connect *it* with tenacity, and to *altogether eliminate* the more simple forms of refutation. One reason for this preference is given by Lakatos himself who has shown that even simple refutations involve a plurality of theories (see especially his paper in *Criticism and the Growth of Knowledge*, 121ff). Another reason is my belief that progress can be brought about only by the active interaction of different 'theories' which of course assumes that the 'research programme' component comes forth not only occasionally, *but is present all the time* (cf. also section 9 below).

[19] It took about twenty-five years before the disturbances of D. C. Miller's repetition of the Michelson–Morley experiment were accounted for in a satisfactory manner. H. A. Lorentz had given up in despair long before that time.

would add that a theory also provides *criteria* of excellence, of failure, of rationality, and that one must support it as long as possible, in order to keep the discourse rational as long as possible. The most important point however is this: it is hardly ever the case that theories are directly compared with 'the facts', or with 'the evidence'. What counts and what does not count as relevant evidence usually depends on the theory *as well* as on other subjects which may conveniently be called 'auxiliary sciences' ('touchstone theories' is Imre Lakatos' apt expression[20]). Such auxiliary sciences may function as additional premises in the derivation of testable statements. But they may also infect the observation language itself, providing the very concepts in terms of which experimental results are expressed. Thus a test of the Copernican view involves on the one hand assumptions concerning the terrestrial atmosphere, the effect of motion upon the objects moved (dynamics); and on the other it also involves assumptions about the relation between sense experience and 'the world' (theories of cognition, theories of telescopic vision included).

The former assumptions function as premises while the latter determine which impressions are veridical and thus enable us not only to *evaluate*, but even to *constitute* our observations. Now there is no guarantee that a fundamental change in our cosmology, such as a change from a geostatic to a heliostatic point of view, will go hand in hand with an improvement of all the relevant auxiliary subjects. Quite the contrary: such a development is extremely unlikely. Who for example would expect the invention of Copernicanism and of the telescope to be at once followed by the appropriate physiological optics: Basic theories and auxiliary subjects are often 'out of phase'. As a result we obtain refuting instances which do not indicate that a new theory is doomed to failure, but only that it does not fit in at present with the rest of science. This being the case scientists must develop methods which permit them to retain their theories in the face of plain and unambiguously refuting facts, even if testable explanations for the clash are not immediately forthcoming. The principle of tenacity (which I call a 'principle' for mnemonic reasons only) is a first step in the construction of such methods.[21]

Having adopted tenacity we can no longer use recalcitrant facts for removing a theory, *T*, even if the facts should happen to be as plain and straightforward as daylight itself. But we can use *other* theories, *T'*, *T''*, *T'''*, etc, which *accentuate* the difficulties of *T* while at the same time promising means for their solution. In this case elimination of *T* is urged by the

[20] Cf. I. Lakatos 'Changes in the Problem of Inductive Logic' in *Mathematics, Science and Epistemology: Philosophical Papers II*, ed. J. Worrall & G. Currie (Cambridge, 1978), 128–93.

[21] For details concerning the 'phase difference' between theories and the corresponding auxiliary sciences, cf. ch. 12 of *A M*. The idea already occurs in Lakatos' *Proofs and Refutations* (Cambridge, 1978); it is a commonplace for Lenin and Trotsky.

principle of tenacity itself.[22] Hence, if change of paradigms is our aim, then we must be prepared to introduce and articulate alternatives to *T* or, as we shall express it (again for mnemonical reasons), we must be prepared to accept a *principle of proliferation.* Proceeding in accordance with such a principle is one method of precipitating revolutions. It is a rational method. Is it the method which science actually uses? Or do scientists stick to their paradigms to the bitter end until disgust, frustration and boredom make it quite impossible for them to go on? What *does* happen at the end of a normal period? We see that our little methodological fairytale makes us indeed look at history with a sharpened vision.

I am sorry to say that I am quite dissatisfied with what Kuhn has to offer on this point. On the one hand he steadfastly emphasizes the dogmatic,[23] authoritarian,[24] and narrowminded[25] features of normal science, the fact that it leads to a temporary 'closing of the mind',[26] that the scientist participating in it 'largely ceases to be an explorer . . . or at least an explorer of the unknown. Instead, he struggles to articulate and concretize the known . . .'[27] so that 'it is [almost always] the individual scientist rather than [the puzzle-solving tradition, or even some particular] current theory which is tested'.[28] 'Only the practitioner is blamed, not his tools.'[29] He realizes of course that a specific science such as physics may contain more than one puzzle-solving tradition, but he emphasizes their 'quasi-independence', asserting that each of them is 'guided by its own paradigms and pursuing its own problems'.[30] A single tradition therefore will be guided by a single paradigm only. This is one side of the story.

On the other hand he points out that puzzle solving is replaced by more 'philosophical' arguments as soon as there exists a choice 'between competing theories'.[31]

Now if normal science is *de facto* as monolithic as Kuhn makes it out to be, then where do the competing theories come from? And if they do arise, then why should Kuhn take them seriously and allow them to bring about a change of the argumentative style, from 'scientific' (puzzle solving) to 'philosophical'?[32] I remember very well how Kuhn criticized Bohm for disturbing the uniformity of the contemporary quantum theory. Bohm's

[22] This is of course not the whole story, but the present sketch suffices entirely for our purpose. Note that Kuhn's argument for tenacity (need for a rational background of argument) is not violated either as the better theory will of course also provide better standards of rationality and excellence.

[23] Kuhn, in *Scientific Change*, 349. [24] *Ibid.*, 393.

[25] *Ibid.*, 350. [26] *Ibid.*, 393.

[27] *Ibid.*, 363. [28] Kuhn, in *Criticism and the Growth of Knowledge*, 5.

[29] *Ibid.*, 7; also cf. Kuhn, *Structure of Scientific Revolutions*, 79.

[30] Kuhn, in *Scientific Change*, 388.

[31] Kuhn, in *Criticism and the Growth of Knowledge*, 7.

[32] 'Philosophical' in Kuhn's (and Popper's) sense and *not* in the sense of, say, contemporary linguistic philosophy.

theory is not permitted to change the argumentative style. Einstein, whom Kuhn mentions in the above quotation, is permitted to do so, perhaps because his theory is now more firmly entrenched than Bohm's. Does this mean that proliferation is permitted as long as the competing alternatives are firmly entrenched? But pre-science which has exactly this feature is regarded as inferior to science. Besides, twentieth-century physics does contain a tradition which wants to isolate the general theory of relativity from the rest of physics, and restrict it to the very large. Why has Kuhn not supported this tradition which is in line with his view of the 'quasi-independence' of simultaneous paradigms? Conversely, if the existence of competing theories involves a change of argumentative style, must we not then doubt this alleged quasi-independence? I have been unable to find a satisfactory answer to these questions in Kuhn's writings.

Let us pursue the point a little further. Kuhn has not only admitted that multiplicity of theories changes the style of argumentation. He has also ascribed a definite function to such multiplicity. He has pointed out more than once,[33] in complete agreement with our brief methodological remarks, that refutations are impossible without the help of alternatives. Moreover, he has described in some detail the magnifying effect which alternatives have upon anomalies and has explained how revolutions are brought about by such a magnification.[34] He has therefore said, in effect, that scientists create revolutions in accordance with our little methodological model and not by relentlessly pursuing one paradigm and suddenly giving up when the problems get too big.

All this leads now at once to difficulty number three, viz. the suspicion that normal or 'mature' science, as described by Kuhn, *is not even a historical fact.*

6. DOES NORMAL SCIENCE EXIST?

Let us recall what we have so far found to be asserted by Kuhn. First, it is asserted that theories cannot be refuted except with the help of alternatives. Secondly, it is asserted that proliferation also plays a historical role in the overthrow of paradigms. Paradigms have been overthrown because of the way in which alternatives have enlarged existing anomalies. Finally, Kuhn has pointed out that anomalies exist *at any point* of the history of a paradigm.[35] The idea that theories are blameless for decades and even

[33] Cf. T. S. Kuhn, 'Measurement in Modern Physical Science', *Isis*, 52 (1961), 161–93 and also my acknowledgement in vol. 1, ch. 5.

[34] A minor disturbance, still accessible to treatment 'can be seen, from another viewpoint, as a counterinstance, and thus as a source of crisis' (Kuhn *Structure of Scientific Revolutions*, 79). 'Copernicus' astronomical proposal . . . *created* an increasing crisis for . . . the paradigm from which it had sprung' (*ibid.*, 74, my italics), 'Paradigms are not corrigible by normal science *at all*' (*ibid.*, 121, my italics).

[35] *Ibid.*, 80ff and 145.

centuries until a big refutation turns up and knocks them out – this idea, he asserts, is nothing but a myth. Now if this is true, then why should we not start proliferating *at once* and *never* allow a purely normal science to come into existence? And is it too much to hope that scientists thought likewise, and that normal periods, if they ever existed, cannot have lasted very long and cannot have extended over large fields either? A brief look at one example, viz. the last century, shows that this seems indeed to be the case.

In the second third of that century there existed at least three different and mutually incompatible paradigms. They were: (1) the mechanical point of view which found expression in astronomy, in the kinetic theory, in the various mechanical models for electrodynamics as well as in the biological sciences, especially in medicine (here the influence of Helmholtz was a decisive factor); (2) the point of view connected with the invention of an independent and phenomenological theory of heat which finally turned out to be inconsistent with mechanics; (3) the point of view implicit in Faraday's and Maxwell's electrodynamics which was developed, and freed from its mechanical concomitants, by Hertz.

Now these different paradigms were far from being 'quasi-independent'. Quite the contrary: it was their *active interaction* which brought about the downfall of classical physics. The troubles leading to the special theory of relativity could not have arisen without the tension that existed between Maxwell's theory on the one hand and Newton's mechanics on the other (Einstein has described the situation in beautifully simple terms in his autobiography; Weyl has given an equally brief, though more technical account in *Raum, Zeit, Materie*; Poincaré exhibits this tension already in 1899, and then again in 1904, in his St Louis lecture). Nor was it possible to use the phenomenon of Brownian motion for a direct refutation of the second law of the phenomenological theory. The kinetic theory had to be introduced from the very start. Here again Einstein, following Boltzmann, led the way. The investigations leading up to the discovery of the quantum of action, to mention yet another example, brought together such different, incompatible, and occasionally even incommensurable disciplines as mechanics (kinetic theory as used in Wien's derivation of his law of radiation), thermodynamics (Boltzmann's principle of the equal distribution of energy over all degrees of freedom) and wave optics and they would have collapsed had the 'quasi-independence' of these subjects been respected by all scientists. Of course not everyone participated in the debate and the great majority may well have continued attending to their 'tiny puzzles'. However if we take seriously what Kuhn himself is teaching then it was not *this* activity that brought about progress, but the activity of the proliferating minority (and of those experimenters who attended to the problems of this minority, and to their strange predictions). And we may ask whether the majority does not continue solving the old puzzles right through the revolutions. But if this is true then Kuhn's account which *temporally separates* periods of proliferation and periods of monism altogether collapses.[36]

7. A PLEA FOR HEDONISM

It seems, then, that the interplay between tenacity and proliferation which we described in our little methodological fairytale is also an essential feature of the actual development of science. It seems that it is not the puzzle-solving activity that is responsible for the growth of our knowledge but the active interplay of various tenaciously held views. Moreover, it is the invention of new ideas and the attempt to secure for them a worthy place in the competition that leads to the overthrow of old and familiar paradigms. Such inventing goes on all the time. Yet is is only during revolutions that the attention turns to it. This change of attention does not reflect any profound structural change (such as e.g. a transition from puzzle solving to philosophical speculation and testing of foundations). It is nothing but a change of interest and of publicity.

This is the picture of science that emerges from our brief analysis. Is it an attractive picture? Does it make the pursuit of science worthwhile? Is the presence of such a discipline, the fact that we have to live with it, study it, understand it, beneficial to us, or is it perhaps liable to corrupt our understanding and diminish our pleasure?

It is very difficult nowadays to approach such questions in the right spirit. What is worthwhile and what is not are to such a large extent determined by the existing institutions and forms of life that we hardly ever arrive at a proper evaluation of these institutions themselves.[37] The sciences especially are surrounded by an aura of excellence which checks any inquiry into their beneficial effect. Phrases such as 'search for the truth', or

[36] It might be objected that the puzzle-solving activity, though not *sufficient* for bringing about a revolution, is certainly *necessary* as it creates the material which eventually leads to trouble: puzzle solving is responsible for some conditions on which scientific progress depends. This objection is refuted by the Presocratics who progressed (their theories did not just *change*, they were also *improved*) without paying the slightest attention to puzzles. Of course, they did not produce the pattern normal science–revolution–normal science–revolution, etc., in which professional stupidity is periodically replaced by philosophical outbursts only to return again at a 'higher level'. However there is no doubt that this is an advantage as it permits us to be openminded all the time and not only in the middle of a catastrophe. Besides, is not 'normal science' full of 'facts' and 'puzzles' which belong, not to the current paradigm, *but to some earlier predecessors*? And is it not also the case that anomalous facts are often *introduced* by the critics of a paradigm, rather than *used by them* as a starting point for criticism? And if that is true, does it not follow that it is proliferation rather than the pattern normalcy–proliferation–normalcy that characterizes science? So that Kuhn's position would be not only methodologically untenable (see the previous section) but also historically false?

[37] Modern analytic philosophers are trying to show that such evaluation is even *logically impossible*. In this they are but the followers of Hegel, except that they lack his knowledge, his comprehensiveness and his wit.

'highest aim of mankind' are liberally used. Undoubtedly they ennoble their object, but they also remove it from the domain of critical discussion (Kuhn has gone one step further in this direction, conferring some dignity even on the most boring and most pedestrian part of the scientific enterprise: normal science). Yet why should a product of human ingenuity be allowed to put an end to the very same questions to which it owes its existence? Why should the existence of this product prevent us from asking the most important question of all, the question to what extent the happiness of individual human beings, and to what extent their freedom, has been increased? Progress has always been achieved by probing well-entrenched and well-founded forms of life with unpopular and unfounded values. This is how man gradually freed himself from fear and from the tyranny of unexamined systems. Our question therefore is: with what values shall we choose to probe the sciences of today?

It seems to me that the happiness and the full development of an individual human being is now, as ever, the highest possible value. This value does not exclude the values which flow from institutionalized forms of life (truth, valour, self-negation, etc.). It rather encourages them *but only* to the extent to which they can contribute to the advance of some individual. What is excluded is the use of institutionalized values for the condemnation, or perhaps even the elimination, of those who prefer to arrange their lives in a different way. What is excluded is the attempt to 'educate' children in a manner that makes them lose their manifold talents so that they become restricted to a narrow domain of thought, action, emotion. Adopting this basic value we want a methodology and a set of institutions which enable us to lose as little as possible of what we are capable of doing and which force us as little as possible to deviate from our natural inclinations.

Now the brief methodological fairytale which we have sketched in section 6 says that a science that tries to develop our ideas and that uses rational means for the elimination of even the most fundamental conjectures must use a principle of tenacity together with a principle of proliferation. It must be allowed to *retain* ideas in the face of difficulties; and it must be allowed to introduce *new* ideas even if the popular views should appear to be fully justified and without blemish. We have also found that actual science, or at least the part of actual science that is responsible for change and for progress, is not very different from the ideal outlined in the fairytale. But this is a happy coincidence indeed! We are now in full agreement with our wishes as expressed above! Proliferation means that there is no need to suppress even the most outlandish product of the human brain. *Everyone may follow his inclinations* and science, conceived as a critical enterprise, will profit from such an activity. Tenacity: this means that one is encouraged not just to follow one's inclinations, but to develop them further, to raise them, with the help of criticism (which involves a comparison with the existing alternatives) to a higher level of articulation *and thereby to raise their defence to a higher level of consciousness*. The interplay between proliferation and tenacity also amounts to the continuation, on a new level, of the biological development of the species and it may even increase the tendency for useful *biological* mutations. It may be the only possible means of preventing our species from stagnation. This I regard as the final and the most important argument against a 'mature' science as described by Kuhn. Such an enterprise is not only ill conceived and non-existent; its defence is also incompatible with a humanitarian outlook.

8. AN ALTERNATIVE: THE LAKATOS MODEL OF SCIENTIFIC CHANGE

Let me now present in its entirety the picture of science which I think should replace Kuhn's account.

This picture is the synthesis of the following two discoveries. First, it contains Popper's discovery that science is advanced by a critical discussion of alternative views. Secondly, it contains Kuhn's discovery of the function of tenacity which he has expressed, mistakenly I think, by postulating tenacious *periods*. The synthesis consists in Lakatos' assertion (which is developed in his own comments on Kuhn) that proliferation and tenacity do not belong to *successive* periods of the history of science, but are always *copresent*.[38]

When speaking of 'discoveries' I do not mean to say that the ideas mentioned are entirely new, or that they now appear in a new form. Quite the contrary: some of these ideas are as old as the hills. The idea that knowledge can be advanced by a struggle of alternative views and that it depends on proliferation was first put forth by the Presocratics (this has been emphasized by Popper himself), and it was developed into a general philosophy by Mill (especially in *On Liberty*). The idea that a struggle of alternatives is decisive for *science*, too, was introduced by Mach (*Erkenntnis und Irrtum*) and Boltzmann (see his *Populärwissenschaftliche Vorlesungen*), mainly under the impact of Darwinism. The need for tenacity was emphasized by those dialectical materialists who objected to extreme 'idealistic' flights of fancy. And the synthesis, finally, is the very essence of dialectical materialism in the form in which it appears in the writings of Engels, Lenin, and Trotsky. Little of this is known to the 'analytic' or 'empiricist' philosophers of today who are still very much under the influence of the Vienna

[38] Lakatos' analysis, I think, can be further improved by abandoning the distinction between theories and research programmes (cf. n.18 of this chapter) and by allowing for incommensurability (jumps from quantity to quality in the language of dialectical materialism). Improved in this way it would be a truly dialectical account of the development of our knowledge.

Circle. Considering this narrow, though quite 'modern' context we may therefore speak of genuine though quite belated, 'discoveries'.

According to Kuhn mature science is a *succession* of normal periods and of revolutions. Normal periods are monistic; scientists try to solve puzzles resulting from the attempt to see the world in terms of a single paradigm. Revolutions are pluralistic until a new paradigm emerges that gains sufficient support to serve as the basis for a new normal period.

This account leaves unanswered the problem of how the transition from a normal period to a revolution is brought about. In section 6 we indicated how the transition could be achieved in a reasonable manner: one compares the central paradigm with alternative theories. Professor Kuhn seems to be of the same opinion. Moreover he points out that this is what actually happens. Proliferation sets in already *before* a revolution and is instrumental in bringing it about. But this means that the original account is faulty. Proliferation does not *start* with a revolution; it *precedes* it. A little imagination and a little more historical research then shows that proliferation not only immediately precedes revolutions, but that it is there *all the time.* Science as we know it is not a temporal succession of normal periods and of periods of proliferation; it is their *juxtaposition.*

Seen in this way the transition from pre-science to science does not *replace* the uninhibited proliferation and the universal criticism of the former by the puzzle-solving tradition of a normal science. It *supplements* it by this activity or, to express it even better, mature science *unites* two very different traditions which are often separate, the tradition of a pluralistic philosophical criticism and a more practical (and less humanitarian; see section 7) tradition which explores the potentialities of a given material (of a theory; of a piece of matter) without being deterred by the difficulties that might arise and without regard to alternative ways of thinking (and acting). We have learned from Professor Popper that the first tradition is closely connected with the cosmology of the Presocratics. The second tradition is best exemplified by the attitude of the members of a closed society towards their basic myth. Kuhn has conjectured that mature science consists in the *succession* of these two different patterns of thought and action. He is right in so far as he has noticed the normal, or conservative, or anti-humanitarian element. This is a genuine discovery. He is wrong in that he has misrepresented the relation of this element to the more philosophical (i.e. critical) procedures. I suggest in accordance with Lakatos' model that the correct relation is one of *simultaneity* and *interaction.* I shall therefore speak of the normal *component* and the philosophical *component* of science and not of the normal *period* and the *period* of revolution.

It seems to me that such an account overcomes many difficulties, both logical and factual, which make Kuhn's point of view so fascinating but at the same time so unsatisfactory.[39] In considering it one should not be misled by the fact that the normal component almost always outweighs its philosophical part. For what we are investigating is not the size of a certain element of science, but its *function* (a single man can revolutionize an epoch). Nor must we be overly impressed by the fact that most scientists would regard the 'philosophical' component as lying outside science proper and that they could support this attitude by pointing to their own lack of philosophical acumen. For it is not they who carry out fundamental improvement but those who further the active interaction of the normal and the philosophical component (this interaction consists almost always in the criticism of what is well entrenched and unphilosophical by what is peripheral and philosophical). Now, granting all this, why is it that there seems to exist a definite fluctuation in the state of science? If science consists of the constant interaction of a normal and a philosophical part; if it is this interaction which advances it; then why do the revolutionary elements become visible only on such rare occasions? Is not this simple historical fact sufficient to support Kuhn's account over mine? Is it not typical philosophical sophistry to deny what is such an obvious historical fact?

I think that the answer to this question is obvious. The normal component is large and well entrenched. Hence, a change of the normal component is very noticeable. So is the resistance of the normal component to change. This resistance becomes especially strong and noticeable in periods where a change seems to be imminent. It is directed against the philosophical component and brings it into public consciousness. The younger generation, always eager for new things, seizes upon the new material and studies it avidly. Journalists, always on the lookout for headlines – the more absurd, the better – publicize the new discoveries (which are those elements of the philosophical component which most radically disagree with the current views while still possessing some plausibility and perhaps even some factual support). These are some reasons for the differences which we perceive. I do not think that one should look for anything more profound.

Now as regards the change of the normal component itself there is no reason to expect that it will follow a clearly recognizable and logical pattern. Kuhn like other philosophers before him (I am here mainly

[39] To take but one example, Kuhn writes (in *Criticism and the Growth of Knowledge*, 6) that 'it is for the normal, not the extraordinary practice of science that professionals are trained; if they are nevertheless eminently successful in displacing and replacing the theories on which normal science depends, that is an oddity which must be explained'. It is certainly an oddity in Kuhn's account. In our account we only need to draw attention to the fact that revolutions are mostly made by members of the philosophical component who, while aware of the normal practice, are also able to think in a different way (in the case of Einstein the self-professed ability to escape from the normal training was essential for his freedom of thought and for his discoveries).

thinking of Hegel) assumes that a tremendous historical change must exhibit a logic of its own and that the change of an idea must be reasonable in the sense that there exists a link between the *fact* of change and the *content* of the idea changing. This is a plausible assumption as long as one is dealing with reasonable people: changes in the *philosophical component* most likely *can* be explained as the result of clear and unambiguous *arguments*. But to assume that people who habitually resist change; who frown at any criticism of things dear to them; and whose highest aim is to solve puzzles on a basis that is neither known nor understood; to assume that *such* people will change their allegiance in a reasonable fashion is carrying optimism and the quest for rationality too far. The normal elements, i.e. those elements which have the support of the majority, may change because the younger generation cannot be bothered to follow their elders; or because some public figure has changed his mind; or because some influential member of the establishment has died and has failed (perhaps because of his suspicious nature) to leave behind a strong and influential school, or because a powerful and non-scientific institution pushes thought in a definite direction.[40] Revolutions, then, are the outward manifestation of a change of the normal component that cannot be accounted for in any reasonable fashion. They are substance for anecdotes though they magnify and make visible the more rational elements of science, thus teaching us what science *could* be if there were more reasonable people around.

9. THE ROLE OF REASON IN SCIENCE

(1) So far I have *criticized* Kuhn from a point of view which is almost identical with that of Lakatos. (There are some slight differences, such as

[40] It is plausible to assume that *one* of the causes for the transition to mature sciences with its various 'quasi-independent' traditions is to be sought in the decree of the Roman Catholic Church against the Copernican point of view. 'This must be taken into account by those who try to explain the special development of the many individual sciences and the absence of a conscious and secure philosophical background by regarding it as a peculiarity of seventeenth-century Italian culture. . . . Such an interpretation assumes . . . that the condemnation of Galileo was but an *external* pressure which could not possibly have influenced the development of spiritual matters. However the Roman Judgement was regarded as a restriction of consciousness that could be broken only on pain of life and salvation. . . . The development of individual disciplines was allowed. Nobody was prevented from searching the heavens, from exploring physical phenomena, from thinking mathematically . . . and from furthering the material culture by such a pursuit. Priests and religious orders, even the Jesuits who were responsible for Galileo's fate, diligently pursued these restricted tasks. But individual conscience as well as the omnipresent 'directeurs de conscience', the officials, the schools, the churches, the state watched carefully this simple fight for knowledge in order that no one might dare to use its results for philosophical speculation'. (Leonardo Olschki *Geschichte der neusprachlichen wissenschaftlichen Literatur* III *Galilei und seine Zeit* (Leipzig, 1927, 400). *This is how 'mature science' came into being*, at least in the Roman countries. Cf. also ch. 9 of A. Wohlwill *Galileo und sein Kampf für die Kopernikanische Lehre* II (Leipzig, 1926), where the development after Galileo's death is sketched in some detail.

my reluctance to separate theories and research programmes,[41] but they will be disregarded. When speaking of 'theories' I always mean theories and/or research programmes.) I now want to *defend* Kuhn against Lakatos. More specifically, I want to argue that science both is, and should be, more irrational than Lakatos and Feyerabend$_1$ (the Popperian$_3$ author of the preceding sections of this paper and of 'Problems of Empiricism') are prepared to admit.[42]

This transition from criticism to defence does not mean that I have changed my mind. Nor can it be completely explained by my cynicism *vis-à-vis* the business of philsosophy of science. It is rather connected with the nature of science itself, with its complexity, with the fact that it has different aspects, that it cannot be readily separated from the remainder of history, that it has always utilized and continues to utilize every talent and every folly of man. Contrary arguments bring out the different features it contains, they challenge us to make a decision, they challenge us to either *accept* this many-faced monster and be devoured by it, or else to *change* it in accordance with our wishes. Let us now see what can be said against the Lakatos model of scientific growth.

(2) Naive falsification judges (i.e. accepts, or condemns) a theory as soon as it is introduced into the discussion. Lakatos gives a theory time, he permits it to develop, he permits it to show its hidden strength, and he judges it only 'in the long run'. The 'critical standards' he employs provide for an interval of hesitation. They are applied 'with hindsight'.[43] They are applied *after* the occurrence of either 'progressive' or of 'degenerating' problem shifts.

Now it is easy to see that standards of this kind have practical force only if they are combined with a *time limit* (what looks like a degenerating problem shift may be the beginning of a much longer period of advance). But introduce the time limit and the argument against naive falsificationism reappears with only a minor modification (if you are permitted to wait, why not wait a little longer?). Thus the standards which Lakatos wants to defend are either *vacuous* – one does not know when to apply them – or they can be *criticized* on grounds very similar to those which led to them in the first place.

In these circumstances one can do one of the following two things. One can stop appealing to permanent standards which remain in force throughout history and govern every single period of scientific development and

[41] Cf. n.18 of this chapter.

[42] The indices are intended as an ironical criticism of Lakatos ('Criticism and the Methodology of Scientific Research Programmes' *Proc. Arist. Soc.*, 69 (1968), 149–86) where the practice of splitting a guy into three was first introduced. (Also cf. Lakatos 181.) This practice has created a lot of confusion and has slowed down philosophers in their attempt to find the weak spots of critical rationalism.

[43] Lakatos in *Criticism and the Growth of Knowledge*, 134, 158, and 173.

every transition from one period to another. Or one can retain such standards as a verbal ornament, as a memorial to happier times when it was still thought possible to run a complex and often catastrophic business like science by following a few simple and 'rational' rules. It seems that Lakatos wants to choose the second alternative.

(3) Choosing the second alternative means abandoning permanent standards *in fact* though retaining them *in words*. *In fact*, Lakatos' position now seems to be identical with the position of Popper as summarized in a marvellous (because self-destructive) addendum of the fourth edition of the *Open Society*.[44] According to Popper we do not 'need any . . . definite frame of reference for our criticism', we may revise even the most fundamental rules and drop the most fundamental demands if the need for a different measure of excellence should arise.[45] Is such a position irrational? Does it imply that science is irrational? *Yes and no*. *Yes*, because there no longer exists a single set of rules that will guide us through all the twists and turns of the history of thought (science), either as participants, or as historians who want to reconstruct its course. One can of course *force* history into such a pattern, but the results will always be poorer and much less interesting than the actual events. *No*, because each particular episode is rational in the sense that some of its features can be explained in terms of reasons which were either accepted at the same time as its occurrence, or invented in the course of its development. *Yes*, because even these logical reasons which change from age to age are never sufficient to explain *all* the important features of a particular episode. We must add accidents, prejudices, material conditions (such as the existence of a particular type of glass in one country and not in another), the vicissitudes of married life, oversight, superficiality, pride, and many other things in order to get a complete picture. *No*, because transported into the climate of the period under consideration and endowed with a lively and curious intelligence we might have had still more to say, we might have tried to overcome accidents, and to 'rationalize' even the most whimsical sequence of events. But – and now we come to a decisive point – how is the transition from certain standards to other standards to be achieved? More especially, what happens to our standards (as opposed to our theories) during a period of revolution? Are they changed in the Popperian manner, by a critical discussion of alternatives, or are there processes which defy a rational analysis? This is one of the questions raised by Kuhn. Let us see what answer we can give to it!

(4) That standards are not always adopted on the basis of argument has been emphasized by Popper himself. Children, he says, 'learn to imitate others . . . and so learn to look upon standards of behaviour as if they

[44] K. R. Popper, 'Facts, Standards, and Truth: a further criticism of relativism', Addendum 1 in the 4th edition of *The Open Society and its Enemies* (Princeton, 1962), II, 388.

[45] *Ibid.*, 390.

consisted of fixed, "given" rules . . . and such things as sympathy and imagination may play an important role in this development'.[46] Similar considerations apply to those grownups who want to continue learning and who are intent on expanding both their knowledge and their sensibility. We certainly cannot assume that what is possible in the case of children – to slide, on the smallest provocation, into entirely new reaction patterns – should be beyond the reach of adults and inaccessible to one of the most outstanding adult activities, science. Moreover, it is likely that catastrophic changes, frequent disappointment and expectations, crises in the development of our knowledge will change and, perhaps, multiply reaction patterns (including patterns of argumentation) just as an ecological crisis multiplies mutations. This may be an entirely natural process, like growing in size, and the only function of rational discourse may consist in increasing the mental tension that precedes *and causes* the behavioural outburst. Now, is this not exactly the kind of change we may expect at periods of scientific revolution? Does it not restrict the effectiveness of arguments (except as a causative agent leading to developments very different from what is demanded by their *content*?) Does not the occurrence of such a change show that science which, after all, is part of the evolution of man is not entirely rational and cannot be entirely rational? For if there are events, not necessarily arguments, which *cause* us to adopt new standards, will it then not be up to the defenders of the status quo to provide, not just arguments, but also *contrary causes*? And if the old forms of argumentation turn out to be too weak a contrary cause, must they then not either give up, or resort to stronger and more 'irrational' means? (It is very difficult, and perhaps entirely impossible, to combat the effects of brainwashing by argument.) Even the most puritanical rationalist will then be forced to leave argument and to use, say, *propaganda*, not because some of his arguments have ceased to be valid, but because the psychological conditions which enable him to effectively argue in this manner and thereby to influence others have disappeared. And what is the use of an argument that leaves people unmoved?

(5) Considering questions such as these a Popperian will reply that new standards may indeed be discovered, invented, accepted, imparted upon others in a very irrational manner, but that there always remains the possibility to criticize them *after* they have been adopted and that it is this possibility which keeps our knowledge rational. 'What, then, are we to trust?' asks Popper after a survey of possible sources for standards.[47] 'What are we to accept? The answer is: whatever we accept we should trust only tentatively, always remembering that we are in possession, at best, of partial truth (or rightness), and that we are bound to make at least some mistake or misjudgement somewhere, not only with respect to facts but also

[46] *Ibid.*, 390. [47] *Ibid.*, 391.

with respect to the adopted standards; secondly, we should trust (even tentatively) our intuition only if it has been arrived at as the result of many attempts to use our imagination; of many mistakes, of many tests, of many doubts, and of searching criticism.'

Now this reference to tests and to criticism which is supposed to guarantee the rationality of science and, perhaps, of our entire life may be either to *well-defined procedures* without which a criticism or test cannot be said to have taken place, or it may be purely *abstract* so that it is left to us to fill it now with this, and now with that concrete content. The first case has just been discussed. In the second case we have but a verbal ornament, just as Lakatos' defence of his own 'objective standards' turned out to be a verbal ornament. The questions of (4) remain unanswered in either case.

(6) In a way even this situation has been described by Popper who says that 'rationalism is necessarily far from comprehensive or self-contained'.[48] But the question raised by Kuhn is not whether *there are* limits to our reason; the question is *where* these limits are *situated*. Are they outside the sciences so that science itself remains entirely rational, or are irrational changes an essential part of even the most rational enterprise that has been invented by man? Does the historical phenomenon 'science' contain ingredients which defy a rational analysis? Can the abstract aim to come closer to the truth be reached in an entirely rational manner, or is it perhaps inaccessible to those who decide to rely on argument only? These are the problems to which we must now address ourselves.

(7) Considering these further problems Popper and Lakatos reject 'mob psychology'[49] and assert the rational character of *all* science. According to Popper it is possible to arrive at a judgement as to which of two theories is closer to the truth, even if the theories should be separated by a catastrophic upheaval such as a scientific revolution. (A theory T' is closer to the truth than another theory, T, if the class of the true consequences of T', the so-called truth content of T', exceeds the class of true consequences of T without an increase in the falsity content.) According to Lakatos the apparently unreasonable features of science occur only in the material world and in the world of (psychological) thought; they are absent from the 'world of ideas, [from] Plato's and Popper's "third world"'.[50] It is in this third world that the growth of knowledge takes place and that a rational judgement of all aspects of science becomes possible. It must be pointed out, however, that the scientist is unfortunately dealing with the world of matter and of (psychological) thought also and that the rules which create order in the third world may be entirely inappropriate for creating order in the brains of living human beings (unless these brains and their structural features are put into the third world, a point that does not become clear

[48] Popper, *Open Society*, ch. 24.

[49] Lakatos, in *Criticism and the Growth of Knowledge*, 178.

[50] *Ibid.*, 180.

from Popper's account).[51] The numerous deviations from the straight path of rationality which we observe in actual science may well be necessary if we want to achieve progress with the brittle and unreliable material (instruments, brains, etc.) at our disposal.

However there is no need to pursue this objection further. There is no need to argue that real science may differ from its third world image *in precisely those respects* which make progress possible.[52] For the Popperian model of an approach to the truth breaks down even if we confine ourselves to ideas entirely. It breaks down because there are *incommensurable theories*.

(8) With the discussion of incommensurability, I come to a point of Kuhn's philosophy which I wholeheartedly accept. I am referring to his assertion that succeeding paradigms can be evaluated only with difficulty and that they may be altogether incomparable, at least as far as the more familiar standards of comparison are concerned (they may be readily comparable in other respects). I do not know which of us was the first to use the term 'incommensurable' in the sense that is at issue here. It occurs in Kuhn's *Structure of Scientific Revolutions* and in my essay 'Explanation, Reduction, and Empiricism' (vol. 1, ch. 5) both of which appeared in 1962. I still remember marvelling at the pre-established harmony that made us not only defend similar ideas but use exactly the same words for expressing them. The coincidence is of course far from mysterious. Both of us had examined the problem before, though in different terms and with somewhat different results (for this 'Prehistory' see my thesis 'Zur Theorie der Basissätze' (Vienna, 1951); vol. 1, ch. 2.6; as well as *SFS*, 65ff). Then I read earlier drafts of Kuhn's book and discussed their content with Kuhn. In these discussions we both agreed that new theories, while often better and more detailed than their predecessors were not always rich enough to deal with *all* the problems to which the predecessor had given a definite and precise answer. The growth of knowledge or, more specifically, the replacement of one comprehensive theory by another involves losses as well as gains. Kuhn was fond of comparing the scientific world view of the seventeenth century with the Aristotelian philosophy, while I used more recent examples such as the theory of relativity and the quantum theory. We also

[51] I am here referring to Popper's 'Epistemology without a Knowing Subject' in *Proceedings of the Third International Congress for Logic, Methodology and Philosophy of Science*, ed. B. Van Rootselaar & S. F. Staal (1968) 333–73 and 'On the Theory of the Objective Mind' in *Proceedings of the Fourteenth International Congress of Philosophy* (1968) I, 25–53. In the first paper birdnests are assigned to the 'Third World' (341) and an interaction is assumed between them and the remaining worlds. They are assigned to the third world *because of their function*. But then stones and rivers can be found in this third world, too, for a bird may sit on a stone, or take a bath in a river. As a matter of fact, everything that is noticed by some organism (and therefore plays a role in his *Umwelt*) will be found in the third world which will therefore contain the whole material world and all the mistakes mankind has made. It will also contain 'mob psychology'.

[52] Cf. *AM*, chs. 13–15.

saw that it might be extremely difficult to compare successive theories in the usual manner, i.e. by an examination of consequence classes. The accepted scheme is as follows (Fig. 1): T is superseded by T'. T' explains why T fails where it does (in F); it also explains why T has been at least partly successful (in S); and makes additional predictions (A). Now if this scheme is to work then there must be statements which follow (with or without the help of definitions and/or correlation hypotheses) both from T and from T'. But there are cases which invite a comparative judgement without satisfying the conditions just stated. The relation between such theories is as shown in Fig. 2.[53] A judgement involving a comparison of content classes is now clearly impossible. For example, T' cannot be said to be either closer to, or farther from, the truth than T.

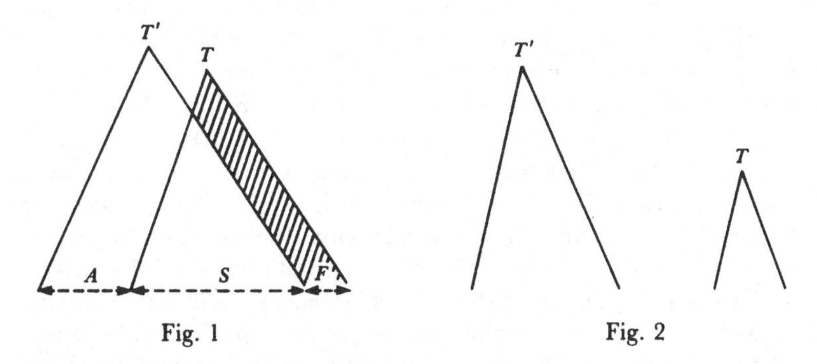

Fig. 1 Fig. 2

(9) As an example of two incommensurable theories let us briefly discuss classical celestial mechanics (CM) and the special theory of relativity (SR). To start with one should emphasize that the question 'are CM and SR incommensurable?' is not a complete question. Theories can be interpreted in different ways. They will be commensurable in some interpretations, incomparable in others. Instrumentalism, for example, makes commensurable all those theories which are related to the same observation language and are interpreted on its basis. Definitions may make some theoretical statements incommensurable; but observation statements can be easily compared. A realist, on the other hand, wants to give a unified account, both of observable and of unobservable matters, and he will use the most abstract terms of whatever theory he is contemplating for that purpose. This is an entirely natural procedure. SR, so one would be inclined to say, does not just invite us to rethink *unobserved* length, mass, duration; it would seem to entail the relational character of *all* lengths, masses, durations, whether observed or unobserved, observable or unobservable. Now extending the concepts of a new theory T to all its

[53] The area below T' should be imagined as lying either *in front of* the area below T, *or behind it*, so that there is no overlap.

consequences, observational reports included, may change the interpretation of these consequences to such an extent that they disappear from the consequence classes of earlier theories. These earlier theories will then all become incommensurable with *T*. The relation between SR and CM is a case in point. The concept of length as used in SR and the concept of length as presupposed in CM are different concepts. Both are *relational* concepts, and very complex relational concepts at that (just consider determination of length in terms of the wavelength of a specified spectral line). But relativistic length (or relativistic *shape*) involves an element that is absent from the classical concept and whose use makes classical concepts inapplicable.[54] It involves the *relative velocity* of the object concerned in some reference system. It is of course true that the relativistic scheme very often gives us *numbers* which are practically identical with the numbers we get from CM; but this does not make the *concepts* more similar. Even the case $c \rightarrow \infty$ (or $v \rightarrow 0$) which gives *strictly identical* predictions cannot be used as an argument for showing that the concepts must coincide at least in this case: different magnitudes based on different concepts may give identical values on their respective scales without ceasing to be different magnitudes (the same remark applies to the attempt to identify classical mass with relative *rest* mass).[55] This conceptual disparity, if taken seriously, infects even the most 'ordinary' situations: the relativistic concept of a certain shape, such as a table, or of a certain temporal sequence, such as my saying 'yes', will differ from the corresponding classical concept also. It is therefore in vain to expect that sufficiently long derivations may eventually return us to the older ideas.[56] The consequence classes of SR and CM are related as in Fig. 2. A comparison of content and a judgement of verisimilitude cannot be made.

(10) In what follows I shall discuss a few objections which have been raised, not against this *particular* analysis of the relation between SR and CM, but against the very *possibility*, or *desirability* of incommensurable theories (almost all objections against incommensurability are of this general kind). They express methodological ideas which we must criticize if we want to increase out freedom *vis-à-vis* the sciences.

One of the most popular objections proceeds from the version of realism

[54] It is possible to base spacetime frames on this new element only and to avoid contamination by earlier modes of thought. Details are given in vol. 1, ch. 7.

Note that the mere *difference* of CM-concepts and SR-concepts does not suffice to make the theories incommensurable. This point is occasionally overlooked by Kuhn. The show must be rigged in such a fashion that the use of any concept of one of the theories makes inapplicable the concepts of the other. Cf. the comments in vol. 1, ch. 1.5, as well as the detailed discussion in ch. 17 of *AM*. *AM*, 269f contains a definition which should, however, be read together with the example that precedes it.

[55] For this point and further arguments, cf. A. S. Eddington, *The Mathematical Theory of Relativity* (Cambridge, 1924), 33.

[56] This takes care of an objection which John Watkins has raised on various occasions.

that I just described in (9). 'A realist', we said, 'wants to give a unified account, both of observable and of unobservable matters, and he will use the most abstract terms of whatever theory he is contemplating for that purpose'. He will use such terms in order to either *give* meaning to observation sentences, or else to *replace* their customary interpretation (e.g. he will use the ideas of S R in order to replace the customary C M-interpretation of everyday statements about shapes, temporal sequences, and so on). As against this it is pointed out that theoretical terms receive their interpretation by being connected either with a pre-existing observation language, or with another theory that has already been connected with such an observation language and that they are devoid of content without such a connection. Thus Carnap asserts[57] that 'there is no independent interpretation for L_T [the language in terms of which a certain theory, or a certain world view, is formulated]. The system T [consisting of the axioms of the theory and the rules of derivation] is itself an uninterpreted postulate system. [Its] terms obtain only an indirect and incomplete interpretation by the fact that some of them are connected by the [correspondence rules] C with observational terms.' Now, if theoretical terms have no 'independent interpretation' then they cannot be used for correcting the interpretation of the observation statements which is the one and only source of their meaning. It follows that realism as described by us is an impossible doctrine.

The guiding idea behind this objection is that new and abstract languages cannot be introduced in a direct way but must be first connected with an already existing, and presumably stable, observational idiom.[58]

This guiding idea is refuted at once by pointing to the way in which children learn to speak and in which anthropologists and linguists learn the unknown language of a newly discovered tribe.

The first example is instructive for other reasons also, for incommensurability plays an important role in the early months of human development. As has been suggested by Piaget and his school,[59] the child's perception develops through various stages before it reaches its relatively stable adult form. In one stage objects seem to behave very much like after-images,[60]

[57] See R. Carnap, 'The Methodological Character of Theoretical Concepts' in *Minnesota Studies in the Philosophy of Science*, ed. H. Feigl & M. Scriven (Minneapolis, 1956) I, 47.

[58] An even more conservative principle is sometimes used when discussing the possibility of languages with a logic different from our own. Thus B. Stroud in 'Conventionalism and the Indeterminacy of Translation', *Synthese*, 18 (1968), 82–96 discussing, and not just stating the principle, says that 'any allegedly new possibility must be capable of being fitted into, or understood in terms of, our present conceptual or linguistic apparatus' from which it follows (172) that 'any "alternative" is either something we already understand and can make sense of, or it is no alternative at all'. What is overlooked is that an initially ununderstood alternative may be *learned* in the way in which one learns a new and unfamiliar language, not by *translation*, but by *living* with the members of the community where the language is spoken.

[59] As an example the reader is invited to consult J. Piaget, *The Construction of Reality in the Child* (New York, 1954).

[60] *Ibid.*, 5ff.

and they are treated as such: the child follows the object with his eyes until it disappears and he does not make the slightest attempt to recover it even if this would require a minimal physical (or intellectual) effort, an effort moreover that is already within the child's reach. There is not even a tendency to search; and this is quite appropriate, 'conceptually' speaking. For it would indeed be nonsensical to 'look for' an after-image. Its 'concept' does not provide for such an operation.

The arrival of the concept, and of the perceptual image, of material objects changes the situation quite dramatically. There occurs a drastic reorientation of behavioural patterns and, so one may conjecture, of thought. After-images or things somewhat like them still exist, but they are now difficult to find and must be discovered by special methods (the earlier visual world therefore *literally disappears*). Such methods proceed from a new conceptual scheme (after-images occur in *humans*, not in the outer physical world, and are tied to them) and cannot lead back to the exact phenomena of the previous stage (these phenomena should therefore be called by a different name, such as 'pseudo-after-images'). Neither after-images, nor pseudo-after-images are given a special position in the new world. For example, they are not treated as *evidence* on which the new notion of a material object is supposed to rest. Nor can they be used to *explain* this notion: after-images arise *together with it* and are absent from the mind of those who do not yet recognize material objects; and pseudo-after-images *disappear* as soon as such recognition takes place. It is to be admitted that every stage possesses a kind of observational 'basis' to which one pays special attention and from which one receives a multitude of suggestions. However this basis (i) changes from stage to stage; and (ii) is part of the conceptual apparatus of a given stage, *not* its one and only source of interpretation.

Considering developments such as these we may suspect that the family of concepts centring upon 'material object' and the family of concepts centring upon 'pseudo-after-images' are incommensurable in precisely the sense that is at issue here. Is it reasonable to expect that conceptual changes of this kind occur only in childhood? Should we welcome the fact – if it is a fact – that an adult is stuck with a stable perceptual world and an accompanying stable conceptual system which he can modify in many ways but whose general outlines have forever become immobilized? Or is it not more realistic to assume that fundamental changes, entailing incommensurability, are still possible, and that they should be encouraged lest we remain forever excluded from what might be a higher stage of knowledge and of consciousness? Besides, the question of the mobility of the adult stage is at any rate an empirical question which must be attacked by *research* and cannot be settled by methodological *fiat*. An attempt to break through the boundaries of a given conceptual system and to escape the range of 'Popperian spectacles'[61] is an essential part of such research.[62]

(11) Looking now at the second element of the refutation – anthropological field work – we see that what is anathema here (and for very good reasons) is still a fundamental principle for the contemporary representatives of the philosophy of the Vienna Circle. According to Carnap, Feigl, Nagel and others, the terms of a theory receive their interpretation, in an indirect fashion, by being related to a different conceptual system which is either an older theory, or an observation language.[63] Older theories, or observation languages, are adopted not because of their theoretical excellence (they cannot possibly be: the older theories are usually refuted). They are adopted because they are 'used by a certain language community as a means of communication'.[64] According to this method, the phrase 'having much larger relativistic mass than . . .' is partially interpreted by first connecting it with some *prerelativistic terms* (classical terms; commonsense terms) which are 'commonly understood' (presumably as the result of previous teaching in connection with crude weighing methods). This is even worse than the once quite popular demand to clarify doubtful points by translating them into Latin. For while Latin was chosen because of its precision and clarity and also because it was conceptually richer than the slowly evolving vulgar idioms, the choice of an observation language or of an older theory as a basis for interpretation is due to the fact that they are 'antecedently understood', it is due to their *popularity*. Besides, if prerelativistic terms which are pretty far removed from reality – especially in view of the fact that they come from an incorrect theory – can be taught ostensively, for example, with the help of crude weighing methods (and we must assume that they can be so taught, or the whole scheme collapses) then why should we not introduce the relativistic terms *directly*, and *without* assistance from the terms of some other idiom? Finally, it is but plain commonsense that the teaching, or the learning, of new and unknown languages must not be

[61] Cf. Lakatos' paper in *Criticism and the Growth of Knowledge*, 179, n.1.

[62] For the condition of research formulated in the last sentence, cf. vol. 1, ch. 5.8. For the role of observation cf. section 7 of the same article. For the application of Piaget's work to physics and, more especially, to the theory of relativity, cf. the appendix of D. Bohm, *The Special Theory of Relativity* (New York, 1965). Bohm and Schumacher have also carried out an analysis of the different informal structures which underlie our theories. One of the main results of their work is that Bohr and Einstein argued from incommensurable points of view. Seen in this way the case of Einstein, Podolsky and Rosen cannot refute the Copenhagen interpretation, and it cannot be refuted by it. The situation is rather that we have two theories, one permitting us to formulate the Einstein–Podolsky–Rosen thought-experiment, the other not providing the machinery necessary for such a formulation so that we must find independent means of deciding which one to adopt. For further comments on this problem, cf. vol. 1, ch. 6.9.

[63] For what follows, cf. also. ch. 3.

[64] Carnap in *Minnesota Studies* I, 40. Cf. also C. G. Hempel, *Philosophy of Natural Science* (Princeton, 1966), 74ff.

contaminated by external material. Linguists remind us that a perfect translation is never possible, even if we use complex contextual definitions. This is one of the reasons for the importance of *field work* where new languages are learned *from scratch* and for the rejection, as inadequate, of any account that relies on (complete, or partial) translation. *Yet just what is anathema in linguistics is now taken for granted by logical empiricists*, a mythical 'observation language' replacing the English of the translators. Let us start field work in this domain also and let us study the language of new theories not in the definition factories of the double language model, but in the company of those metaphysicians, experimenters, theoreticians, playwrights, courtesans, who have constructed new world views! This finishes our discussion of the guiding principle of the first objection against realism and the possibility of incommensurable theories.

(12) Next I shall deal with a mixed bag of asides which have never been presented in a systematic fashion and which can be disposed of in a few words.

To start with, there is the suspicion that observations which are interpreted in terms of a new theory can no longer be used to refute that theory. The suspicion is allayed by pointing out that the predictions of a theory depend on its postulates, the associated grammatical rules *as well as* on initial conditions, while the meaning of the primitive notions depends on the postulates (and the associated grammatical rules) only: it is possible to refute a theory by an experience that is entirely interpreted in its terms.

Another point that is often made is that there exist *crucial experiments* which refute one or two allegedly incommensurable theories and confirm the other, for example: the Michelson–Morley experiment, the variation of the mass of elementary particles, the transversal Doppler effect refute CM and confirm SR. The answer to this problem is not difficult either: adopting the point of view of relativity we find that the experiments *which of course will now be described in relativistic terms*, using the relativistic notions of length, duration, speed, and so on,[65] are relevant to the theory and we shall also find that they support the theory. Adopting CM (with, or without an ether) we again find that the experiments (which are now described in the very different terms of classical physics, roughly in the manner in which Lorentz described them) are relevant, but we also find that they *undermine* (the conjunction of classical electrodynamics and of) CM. Why should it be necessary to possess terminology that allows us to say that it is the *same* experiment which confirms one theory and refutes the other? But did we not ourselves use such terminology? Well, for one thing it should be easy, though somewhat laborious, to express what was just said *without* asserting identity. Secondly, the identification is of course not contrary to our thesis,

[65] For examples of such descriptions cf. D. L. Synge 'Introduction to General Relativity' in *Relativity, Groups and Topolgy*, ed. B. DeWitt & T. R. DeWitt (New York, 1964).

for we are now not *using* the terms of either relativity, or of classical physics, as is done in a test, but are *referring to* them and their relation to the physical world. The language in which *this* discourse is carried out can be classical, or relativistic, or ordinary. It is no good insisting that scientists act as if the situation were much less complicated. If they act that way, then they are either instrumentalists (see (9) above) or mistaken: many scientists are nowadays interested in *formulae* while we are discussing *interpretations*. It is also possible that being well acquainted with both C M and S R they change back and forth between these theories with such speed that they seem to remain within a single domain of discourse.

(13) It is also said that in admitting incommensurability into science we can no longer decide whether a new view explains what it is supposed to explain or whether it does not wander off into different fields. For example, we would not know whether a newly invented physical theory is still dealing with problems of space and time or whether its author has not by mistake made a biological assertion. But there is no need to possess such knowledge. For once the fact of incommensurability has been admitted the question which underlies the objection does not arise (conceptual progress often makes it impossible to ask certain questions; thus we can no longer ask for the absolute velocity of an object – at least as long as we take relativity seriously). Yet is this not a serious loss for science? Not at all! Progress was made by the very same 'wandering off into different fields' whose undecidability now so greatly exercises the critic: Aristotle saw the world as a super*organism*, that is, as a *biological* entity, while one essential element of the new science of Descartes, Galileo, and of their followers in medicine and in biology is its exclusively *mechanistic* outlook. Are such developments to be forbidden? And if they are not, then what is left of the complaint?

A closely connected objection starts from the notion of *explanation*, or *reduction*, and emphasizes that this notion presupposes continuity of concepts (other notions could be used for starting exactly the same kind of argument). Now to take our above example, relativity is supposed to explain the valid parts of classical physics, hence it cannot be incommensurable with it! The reply is again obvious. Why should the relativist be concerned with the fate of classical mechanics except as part of a historical exercise? There is only *one* task we can legitimately demand of a theory and it is that it should give us a correct account of the *world*. What have the principles of explanation got to do with this demand? Is it not reasonable to assume that a point of view such as that of classical mechanics which has been found wanting in various respects cannot have entirely adequate concepts, and is it not equally reasonable to try replacing its concepts by those of a more successful cosmology? Besides, why should the notion of explanation be burdened by the demand for conceptual continuity? This notion has been found to be too narrow before (demand of derivability) and it had to be widened so as to include partial and statistical connections. Nothing prevents us from widening it still further to admit, say, 'explanation by equivocation'.

(14) Incommensurable theories, then, can be refuted by reference to their own respective kinds of experience (in the absence of commensurable alternatives these refutations are quite weak, however).[66] Their content cannot be compared. Nor is it possible to make a judgement of verisimilitude except within the confines of a particular theory. None of the methods which Popper wants to use for rationalizing science can be applied and the one that can be applied, refutation, is greatly reduced in strength. What remains are aesthetic judgements, judgements of taste, and our own subjective wishes. Does this mean that we are ending up in subjectivism? Does this mean that science has become arbitrary, that it has become one element of the general relativism which Popper wants to attack? Let us see.

To start with, it seems to me that an enterprise whose human character can be seen by all is preferable to one that looks 'objective', and impervious to human actions and wishes.[67] The sciences, after all, are our own creation, including all the severe standards they seem to impose upon us. It is good to be constantly reminded of this fact. It is good to be constantly reminded of the fact that science as we know it today is not inescapable and that we may construct a world in which it plays no role whatever (such a world, I venture to suggest, would be more pleasant than the world we live in today). What better reminder is there than the realization that the choice between theories which are sufficiently general to provide us with a comprehensive world view and which are empirically disconnected may become a matter of taste? That the choice of our basic cosmology may become a matter of taste?

Secondly, matters of taste are not completely beyond the reach of argument. Poems, for example, can be compared in grammar, sound structure, imagery, rhythm, and can be evaluated on such a basis (cf. Ezra Pound on progress in poetry).[68] Even the most elusive mood can be analysed, *and must be* analysed if the purpose is to present it in a manner that can either be enjoyed, or that increases the emotional (cognitive, perceptual) inventory of the reader. Every poet who is not completely irrational compares, improves, argues until he finds the correct formulation of what he wants to

[66] For this point cf. vol. 1, ch. 6.1.

[67] For this problem of 'alienation' cf. Karl Marx 'Nationalökonomie und Philosophie' quoted from *Die Fruehschriften* (Stuttgart, 1953) and 'Zur Kritik der Hegelschen Rechtsphilosophie', *Deutsch-Französische Jahrbücher* (1844).

[68] Popper has repeatedly asserted, both in his lectures and in his writings, that while there is progress in the sciences there is no progress in the arts. He bases his assertion on the belief that the content of succeeding theories can be compared and that a judgement of verisimilitude can be made. The refutation of this belief eliminates an important difference (and perhaps the *only* important differences) between science and the arts and makes it possible to speak of styles and preferences in the first, and of progress in the second.

say.[69] Would it not be marvellous if this process played a role in the sciences also?

Finally, there are more pedestrian ways of explaining the same matter which may be somewhat less repulsive to the ears of a professional philosopher of science. We may consider the *length* of derivations leading from the principles of a theory to its observation language, and we may also draw attention to the number of *approximations* made in the course of the derivation (all derivations must be standardized for this purpose so that an unambiguous judgement of length can be made; this standardization concerns the *form* of the derivation, it does not concern the *content* of the concepts used). Smaller length and smaller number of approximations would seem to be preferable. It is not easy to see how this requirement can be made compatible with the demand for simplicity and generality which, so it seems, would tend to increase both parameters. However that may be, there are many ways open to us once the fact of incommensurability is understood, and taken seriously.

(15) I started by pointing out that scientific method, as softened up by Lakatos, is but an ornament which makes us forget that a position of 'anything goes' has in fact been adopted. I then considered the argument that the method of problemshifts, while perhaps useless in the first world might still give a correct account of what goes on in the third world and that it might permit us to view the whole 'third world' through 'Popperian spectacles'. The reply was that there is trouble in the third world also and that the attempt to judge cosmologies by their content may have to be given up. Such a development, far from being undesirable, changes science from a stern and demanding mistress into an attractive and yielding courtesan who tries to anticipate every wish of her lover. Of course, it is up to us to choose either a dragon or a pussy cat for our company. I do not think I need to explain my own preferences.

[69] Cf. B. Brecht, *Uber Lyrik* (Frankfurt, 1964). In my lectures on the theory of knowledge I usually present and discuss the thesis that finding a new theory for given facts is like finding a new production for a well-known play. For painting, cf. also E. Gombrich, *Art* and Illusion (Princeton, 1960).

APPENDIX

REALISM AND THE BOHR–ROSENFELD CONDITION

1. The idea that the basic concepts of a theory should not be contaminated by extraneous material but should be explained by the theory itself is found not only among professional philosophers. It was responsible for the transition from a qualitative physics that regarded mathematical theories as mere instruments, to a point of view where reality was viewed in mathematical terms and qualities ceased to be of interest for science. More recently Bohr and Rosenfeld[1] and, following them, Marzke and Wheeler[2] have suggested basing the general theory of relativity on relativistic notions alone. 'Why should it be necessary' write the latter authors 'to lean at all upon the atomic constitution of matter to define a standard of length?' (48) 'Why bring the quantum of action into the foundations of classical general relativity? Why found length measurements upon either $\hbar^2/mc^2$ or $\hbar^3c/mc^4$? . . . Classical general relativity should admit to calibrations of spacetime that are altogether free of any reference to the quantum of action.' Let us now see how this suggestion relates to the more general philosophical discussion of the interpretation of theories.

2. In empirical science we may distinguish between complete theories and incomplete theories. Complete theories not only give an account, in their own terms, of all the events in their domain, they are also rich enough to give us means for testing statements about these events. Incomplete theories, on the other hand, must use external ideas in at least some tests. And as the tests of a theory elucidate its cognitive meaning we may say that the meaning of incomplete theories depends on elements that are external to them.

Examples of incomplete theories are the phlogiston theory of calcination and combustion (the weight of substances is determined by procedures that are not defined by the theory itself), the theory of evolution (the shape of organisms is determined by procedures that are not defined by the theory),

[1] 'Zur Frage der Messbarkeit der elektromagnetischen Feldgrössen' *K. danske Vidensk. Selsk. Mat.-fys. Meddr* 12, (1937), 1; *Phys. Rev.* 78 (1950), 794.

[2] 'Gravitation and Geometry I: The Geometry of Space–Time and Geometrodynamical Standard Meter' in *Gravitation and Relativity*, ed. H.-Y. Chiu & W. F. Hoffmann (New York, 1964), 40ff.

the theory of cognitive dissonance and so on. All low level laws (the law of free fall; Kepler's laws; the laws of geometrical optics) are incomplete in the sense just described and so is the quantum theory at least as long as we adopt Bohr's interpretation of it (experimental results are formulated in classical terms). Marzke and Wheeler suggest that classical (i.e. non-quantal) general relativity is a complete theory.

3. Complete theories may be contaminated by external material. Thus the double language model makes every theory depend on a conceptual system (an 'observation language') whose test procedures are set up independently of the assertions made by the theory. Another example is the notion of mass in Newton's mechanics. Mass was here defined as 'amount of matter', independently of inertia and resistance. Mach's definition of mass removed this external element. Even modern scientific theories are often contaminated by common terms, or by terms belonging to an earlier stage of theorizing, or by terms taken from a different and perhaps equally abstract theory which remain because not every scientist is interested in conceptual purity. An example is the general theory of relativity where time intervals are defined by atomic clocks while space measurements still rely (conceptually) on familiar triangulation methods. It may not be possible to disentangle the mixture and make part of it stand on its own feet. But whenever it is, such a disentanglement gives us interesting information about the world that we inhabit and our knowledge of it.

4. Complete theories have much in common with metaphysical world views. They construct an entire world out of a few self-sufficient processes. Accidents of origin often introduce these processes in a misleading form. Thus when measuring distance with the help of rigid rods, and time with the help of inertial clocks the constancy of the velocity of light in all inertial systems becomes a rather puzzling empirical fact. This still reflects the standpoint of the ether theories where we have a basic spacetime framework filled by a complex classical system – the ether – assumed to obey certain basic equations. Knowledge of ether processes and of the interaction between the ether and material systems can be obtained partly by observation, partly by theoretical developments from the basic equations. Proceeding in this way Lorentz explained the constancy of c as an approximate result of a complex balancing process involving opposing effects.[3] Einstein inversed the procedure and made the propagation of light in vacuum a fundamental process. According to Einstein this process is not the *result* of complex interactions; it is a basic feature common to all interactions. This suggests that we either *replace* the old method of time and space measurements by new methods, or regard them as practicable but

[3] Cf. his description of the matter in *The Theory of Electrons* (New York, 1952), 230.

only approximate and conceptually opaque *substitutes* of such new methods. Marzke and Wheeler give an outline of what such new methods will look like.

5. First, local curvature of space is determined by the world lines of particles that have been started off with a specified separation and zero relative velocity. The amount by which the separation decreases gives an estimate of the local curvature. To trace out ideal geodesics, independent of friction, radiation pressure, and electric fields a spherical satellite A is surrounded by a hollow sphere B equipped with rockets R and internal sensors which keep A in the centre of B (fig. A1). Geodesics are needed in addition to light rays because spaces with different geometries may contain exactly the same patterns of light rays and of their intersections. Local flatness can be assumed as long as $L^2/2R^2 <$ measuring error, where R is the radius of curvature and L the maximum length involved in the measurements.

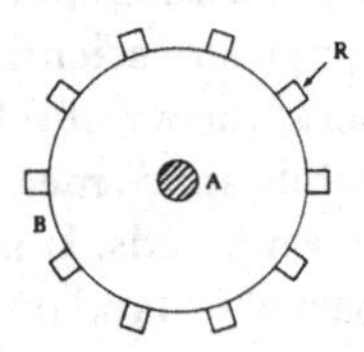

Fig. A1

Having ascertained *local flatness* one constructs *parallels* (51) and *geodesic clocks*: a clock consists of a light ray reflected back and forth between two parallels. Intervals are compared with standard intervals in the following fashion. Assuming AB to be the standard interval and CD the interval to be measured, we connect A with C by a world line W and construct a geodesic clock between W and a close parallel W'. B' and B'' are the events of sending a light ray to B and receiving the light ray reflected from B (fig. A2). Putting $c=1$ we obtain (with τ the time for a single round trip between W and W')

$$\begin{aligned} AB^2 &= (t_B - t_A)^2 - (x_B - x_A)^2 \\ &= (t - 0)^2 - (x - 0)^2 \\ &= t^2 - x^2 = (t + x)(t - x) \end{aligned}$$

Let

$$t + x = N_2\tau \quad t - x = N_1\tau$$

$$AB^2 = N_1N_2\tau^2$$

Similarly

$$CD^2 = N_3N_4\tau^2 \quad \text{and} \quad \frac{AB}{CD} = \sqrt{\frac{N_1N_2}{N_3N_4}}$$

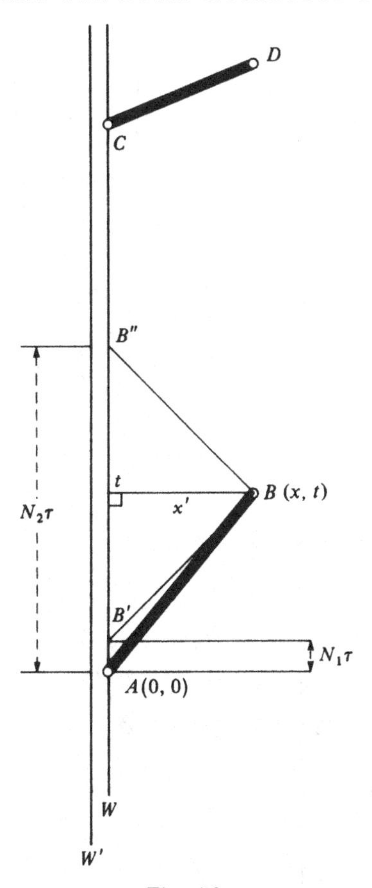

Fig. A2

Velocities are best introduced with the help of the *k*-calculus which directly uses the constancy of the velocity of light and its independence of the motion of the light source.[4] The factor k relates time intervals in inertial systems: the time interval T, measured by a light clock in A is related to the time interval T' cut off on B by light rays emanating from A by the factor k: $T'=kT$. The equivalence of A and B gives $T''=KT'=k^2T$ (see fig. A3). The independence of light of the velocity of the source gives $k_{AC}=k_{AB}.k_{BC}$ (see fig. A4). Distance measurement by the radar method (with $c=1$) gives (see fig. A5) $x=\frac{1}{2}k^2T-T=\frac{1}{2}T(k^2-1)$. This distance is achieved in time T plus the time assigned to event E which is half the round trip time of light from the

[4] The *k*-calculus was introduced by H. Bondi. See his *Assumption and Myth in Physical Theory* (Cambridge, 1967).

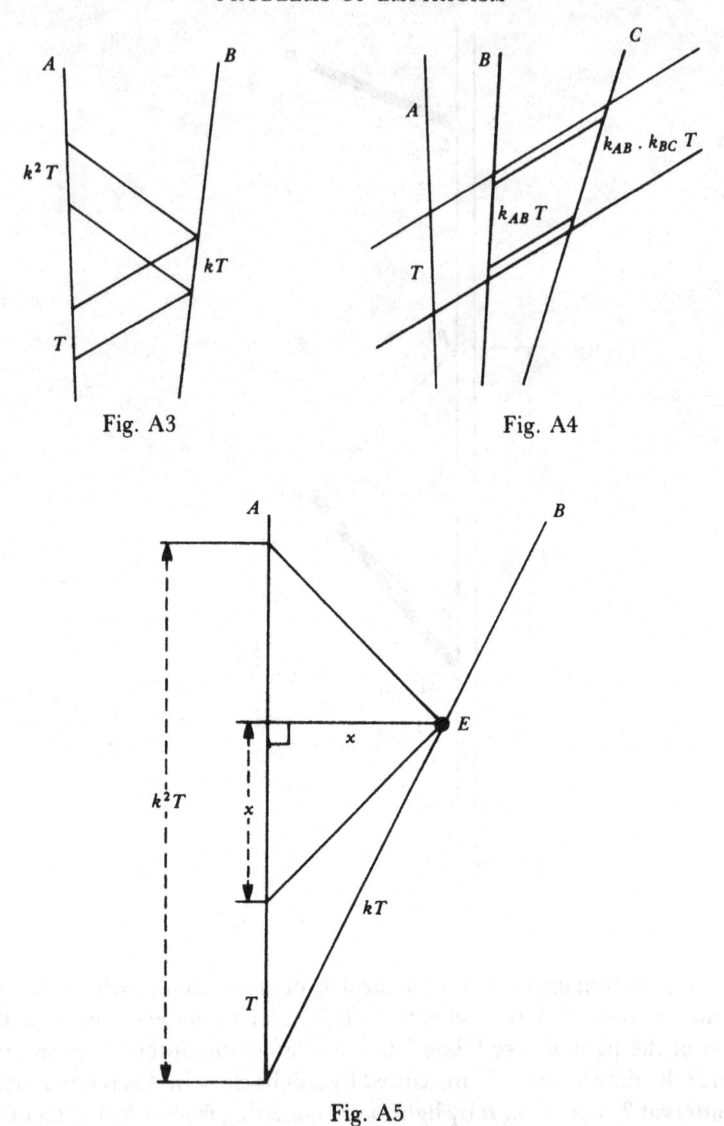

Fig. A5

base to E, i.e. $\frac{1}{2}(k^2 - 1)T$ which gives $\frac{1}{2}T(k^2 + 1)$ so that $v = (k^2 - 1)/(k^2 + 1)$, or via the relation $k_{AC} = k_{AB}.k_{BC}$

$$v_{AC} = (v_{AB} + v_{BC})/(1 + v_{AB}v_{BC})$$

the relativistic addition theorem for velocities.

6. Does the relation between relativity as explained here and classical mechanics make the two theories incommensurable? The question cannot be answered until we have introduced all relativistic concepts. For those mentioned the situation is clear: velocity and distance are defined under the assumption that there exists an emission-independent process whose velocity is the same in all inertial systems. If this assumption is denied then the definitions given and the theorems derived cease to make sense, exactly as is to be expected in the case of incommensurable theories. One can redefine the notions in a manner that assumes that a process of the kind described exists but is the result of a complex interaction of more fundamental processes. The relations will then all be retained but they will be apparent motions (velocities, distances) not real motions.

9
Popper's *Objective Knowledge*

1. CONTENT OF THE BOOK; MAIN THESES

The book[1] is a collection of articles, some new, some reprinted from earlier works. It contains proposals for three areas of research: ontology, methodology, evolution. In ontology Popper proposes a pluralistic approach; in methodology he proposes critical discussion of competing views as a means of improving knowledge; in evolution he proposes a theory that entails a 'Darwinian' (261) epistemology and explains knowledge as an ever-changing 'exosomatic' (251) product of the organism, as a kind of secretion that is constantly modified and augmented by trial and error procedures and that protects the organism from being modified itself. According to this theory knowledge is neither a special form of *belief* – for example, it is not a well-established belief, or a highly probable belief; it is not even a falsifiable and highly corroborated belief – nor does it consist in eternal ideas which we grasp but cannot change. Knowledge is a product of man, it can be changed by man, but is still objective and even autonomous, i.e. it cannot be reduced to either physical or mental processes. It is objective because it obeys laws of its own that are independent of the intentions of its creators: having been produced by man it no longer obeys all his wishes. And it is autonomous because these laws are neither physical laws, nor mental laws, nor reducible to physical and/or mental laws. The phenomenon of knowledge shows that the physical world is an open world and that some of its inhabitants are affected by physical as well as by non-physical influences. This is explained in the following way.

2. ONTOLOGICAL PLURALISM; CRITERIA FOR AUTONOMY; THE THREE WORLDS

According to Popper

> [t]here are many sorts of real things . . . foodstuffs . . . or more resistant objects . . . like stones, and trees, and humans. But there are many sorts of

[1] Karl R. Popper, *Objective Knowledge. An Evolutionary Approach* (Oxford, 1972); reprinted with corrections 1973. (Numbers in parentheses refer to pages of this work unless otherwise indicated.) An earlier version was read and criticized by Imre Lakatos. I have adopted most of his suggestions.

> reality which are quite different, such as our subjective decoding of our experiences of foodstuffs, stones, and trees, and human bodies. . . . Examples of other sorts in this many-sorted universe are: a toothache, a word, a language, a highway code, a novel, a governmental decision; a valid or invalid proof; perhaps forces, fields of forces . . . structures . . . [37]

and, so one might add, ghosts, numbers, spirits, gods, God and the Devil (cf. 153).

Contemplating this great variety of things and processes Popper is led 'to assert that realism', the doctrine that the world exists independently of our perception of it, 'should be at least tentatively pluralistic' (294; cf. 252). A rationalist will of course 'wish and hope to understand the world' that is, he will wish and hope to be able to reduce some of its entities to others (292; cf. 263). Ideally, he will wish to explain everything by the properties of, and the relations between, a few basic processes. But this wish should be permitted to have its way only *after* the plurality of the things to be explained has been recognized: '[it is] only *after* recognizing the plurality of what there is in this world [that we can] seriously begin to apply Ockham's razor' (301, original italics). For it is 'quite likely that there may be no reduction possible'; for example, 'it is conceivable that life is an *emergent* property of physical bodies' (292, original italics).

Taken literally the advice is superfluous. Whoever wants to apply Ockham's razor is aware of 'the plurality of what there is'. As a matter of fact, he is so annoyed by some entities that he wants to cut them off. This he is not supposed to do unless he has first assured himself that he is not eliminating irreducible entities. But how else can he assure himself, according to Popper's own philosophy, than by resolutely trying out monistic programmes? In the past Popper has supported such programmes. The bolder the programme, the higher the praise it received (cf. his paper on the Presocratics in *CR*).[2] Now he is more cautious.[3] Let us see where his caution leads him.

According to Popper there are two criteria which help us to recognize emergent, or irreducible, or autonomous entities.

We recognize their reality by the fact that they *interact* with other entities, and especially with physical objects: '. . . I propose to say that something exists, or is real, if and only if it can *interact* with . . . hard physical bodies' (*E*, 23).

We recognize that they are *irreducible*, or *autonomous* (relative to a given set of entities, or a 'world') from the fact that they (1) have special properties and/or obey special laws not found in the world to which reference is made

[2] *CR* refers to *Conjectures and Refutations* (London, 1963), *E* to Popper's article in *Encounter*, April 1973.

[3] Another instance of this cautiousness is discussed in my article 'In Defence of Classical Physics', *J. Hist. Phil. Sci.* 1(1970), 59ff.

(*CR*, 298, concerning the difference between the laws of physics and the laws of logic; cf. 147 and 225); and (2) that these properties and/or laws are 'unpredictable in principle', given the properties, the laws, and certain arrangements of entities in the reference world (298).

Briefly, but not at all incorrectly: entities are *real* if they can interact with the physical world (world 1 is 'the very standard of reality': *E*, 21). They are *autonomous* (with respect to the physical world) if they are not physical entities.

Note that autonomy as explained here does not mean independence. Most of the autonomous entities and worlds Popper discusses influence each other in complex ways. Some of the influences are even used as arguments for autonomy, as we shall see (cf. 155). Autonomy of a thing, or of a world of things, just means that the thing, or the world, has specific properties and enters into specific relations which cannot be reduced to the properties of, and the relations between, the entities of some other world.

Using his criteria Popper tries to isolate three different types of entities forming three different types of world: the world of physical objects, or world 1; the world of mental processes, or world 2; and the world of theories, arguments, problems, complaints, works of art, or world 3 (it is not implied that these are the only worlds in existence, nor is it assumed that the order in which they are enumerated is the only possible order (106–7)). Much space is devoted to the description of these three worlds, of their interaction, and of the arguments for their autonomy.

I now start with a discussion of what Popper has to say about his world 2.

3. MENTAL PROCESSES

Popper does not always insist on the autonomy of mental processes. Occasionally he concedes the possibility of a reduction: 'should they [the mental processes] one day be reduced to physical [processes], then this will be a tremendous success' (293), but he refuses to discuss the matter further (298: 'Whether . . . mental states . . . may . . . be reduced to physical states . . . will not be further discussed'). On one occasion he even accepts a physicalistic solution of the mind–body problem. Having conjectured 'that in some organisms mental states may *interact* with physical states' (231n) and added that 'the acceptance of an "*interaction*" of mental and physical states offers the only satisfactory solution of Descartes's problem' (252n), he writes: 'The Stoics were materialists; they regarded the soul as part of the body, identifying it with "the breath of life" . . . This theory may, however, be interpreted as a special form of body–mind dualism' (157n). Here a particular form of body–mind *monism* is accepted because it preserves the *apparent* autonomy of world 2 (by making it a quasi-autonomous part of world 1). Most of the time, however, mental events are regarded as irreducible inhabitants of an autonomous world, viz. world 2.

To defend this assumption against materialists or physicalists Popper uses heavy rhetoric and one argument.

The *rhetoric* boils down to the insinuation (already present in *CR*) that the physicalist denies the *reality* and not just the autonomy of mental states: '. . . to show the reality of world 2 one can appeal to commonsense and to the failure of the physicalist to produce telling arguments against the commonsense view that a bad toothache can be very real indeed' (*E*, 21). This is a most naive remark. Where is the physicalist (the *intelligent* physicalist: every half-way popular theory has of course lots of idiots among its defenders) who denies the existence of pain? For the physicalist a pain is as real as a stone, or a planet, or a rainbow, or a thunderstorm. But while his opponent has often believed that the different behaviour, or the different appearance of these things indicates a difference in nature, 'in what may be called [their] ontological status' (161) and has subdivided the universe into relatively autonomous regions, or 'worlds' (superlunar world versus sublunar world; world of man versus the world of the gods; world 1 versus world 2), the physicalist from the very beginning has tried to show how the *same* entities (e.g. atoms) obeying the *same* basic laws give rise to a great variety of phenomena when subjected to changing *conditions* (which consist of further entities of the same kind, obeying the same basic laws, but arranged in a great variety of ways). Turning his attention to mental states he has formed 'tentative theories' (293) to the effect that they, too, are complex physical processes (in the central nervous system), and he has tried to discover the detailed structure of these processes rather than denying their existence. Poppers comments on reduction (292ff) indicate that he does not object to such a procedure. The only thing he objects to in *this* passage are either *ad hoc* reductions which replace mental processes by *unspecified* physical processes *and leave it at that*, or simple eliminations which 'reject [the] existence [of mental states] by merely noting that we can explain things without them' (293). No contemporary materialist would be satisfied with such simpleminded procedures.

The *argument* may be stated as follows.

Human knowledge, including plans, problems, theories, solutions of problems is 'something quite abstract' (230). Abstract entities (numbers, concepts, etc.) cannot have a causal influence upon physical processes (155). Yet human knowledge has transformed the physical world. There must therefore exist processes that mediate between world 3 and world 1. These processes can be neither abstract, nor material. But they must be capable of acting on world 1 and being acted upon by world 3. Now we know that 'we must normally grasp, or understand a world 3 theory before we can use it to act upon world 1' (*E*, 21). Grasping, thinking, understanding seem to be the mediating processes we are looking for. It follows that they cannot be material processes but must form an autonomous domain between world 1 and world 3. This domain is world 2: world 2 cannot be reduced to world 1. So far Popper's 'main argument' (*E*, 21) for the autonomy of world 2.

The following comments should be made.

(1) The argument involves world 3 in an essential way. A discussion of world 1 and world 2 may give rise to various difficulties, just as does any attempt at a reduction; but it does not lead to a single decisive objection against materialism. At least Popper has not managed to produce such an objection. It is the discovery of human knowledge that makes us realize that the physical world is an open system in the sense that it can be changed by non-physical influences, and that mental processes are among such influences (*E*, 26).

(2) The argument shows autonomy only for mental processes involving third-world objects. Pains, feelings, sensations are outside its domain of application (unless it turns out that they too contain an 'intellectual' component).

(3) The argument assumes the autonomy of world 3. It is therefore in need of further arguments which defend that autonomy against monistic research programmes such as materialism.

4. THE AUTONOMY OF WORLD 3

At first sight Popper seems to offer a great variety of such arguments. Closer analysis reveals that he hardly ever argues for autonomy in the sense of section 2, though he likes to give the impression that he does. Some arguments assert the autonomy of world 3 while all that has been shown is its reality, or its (relative) causal independence. Occasionally this is due to the fact that the two ideas – reality and autonomy – are not kept apart,[4] while on other occasions we have simply a bad argument.[5] Some so-called

[4] There are passages where the two ideas are treated as synonyms: 'I suggest *that it is possible to accept the reality or (as it may be called) the autonomy of the third world* . . .' (159, original italics). Autonomy may also simply mean not having been made by man. Mountains and human genes (which have a decisive influence on the structure of our knowledge) are autonomous in this sense. On the other hand, there are passages where the distinction is clear enough, and in agreement with the criterion of section 2: 'there exist objects which have not taken up either world 1 shape or world 2 shape but which, nevertheless, interact with our thought processes', our behaviour, and with the remaining physical world (*E*, 22; 74). '[T]here are many theories in themselves and arguments in themselves and problem situations in themselves which have never been produced or understood and may never be produced or understood by men' (116). Such objects would indeed be autonomous both with respect to world 1 and with respect to world 2. For it is difficult to see how mindless and bodyless causes can be reduced to mental and/or physical events. Popper uses these ambiguities to the full to make a weaker argument (for reality, or physical independence) seem to support a stronger case.

[5] This is true of the 'standard argument for the . . . *independent existence of the third world*' (107, original italics). The argument shows that certain entities (books, libraries, knowledge) are

arguments are not arguments at all but involved theses, or insinuations[6] and even in those rare cases where we deal with genuine and relevant reasons, the reasons are so well embedded in the presentation of the entire three world view and so closely connected with the reality, and even the autonomy of other worlds, that it is difficult to separate them *and* to preserve their strength.[7] There are only two considerations, one confused, but promising, the other clear but much less promising, which cannot be immediately shown to be inadequate.

5. ZOOLOGICAL ANALOGUES

The first consideration is developed by examining certain similarities between the world of human culture and the changes which animals effect in their surroundings.

Popper emphasizes that 'there are only fairly remote analogues in the

necessary for a civilization, without demonstrating their autonomy. It compares two situations: (1) machines, subjective knowledge are destroyed, but books and our capacity for learning are retained; (2) machines and subjective knowledge are destroyed *together with* our libraries, but our capacity for learning is retained. In the first case rebuilding is possible. In the second case 'there will be no re-emergence of our civilization for many millennia' (108). For the materialist this is a consequence of the hypothesis, accepted by Popper, that the development of properly programmed computers by an evolutionary process takes considerable time, and is in any case highly unlikely ever to happen again, taken together with the triviality that computers without programmes don't get very far. Both the hypothesis and the triviality are acceptable to a materialist.

[6] Insinuations masquerading as arguments occur in the defence of the thesis that 'no causal theory of the descriptive and argumentative function of language is possible' (*CR*, 301). Part of this thesis is defended by the remark that a causal theory of *naming* is not possible because the ends of the naming relation (the object and the name) stand out only in our *interpretation* (*CR*, 297). The remark assumes that we cannot give a physical account of this kind of interpretation. That is the assumption at issue. (It is also a false assumption: any mechanism capable of pattern recognition can select the desired pairs of entities from a causal sequence.) Another 'argument' consists in the assertion that insight into the mechanism of a computer 'teaches us' that it does not argue, even if initially we mistakenly assumed that it did (*CR*, 296): mechanisms don't argue because they don't argue, only not expressed in such a straightforward manner.

[7] We have already met an example of this situation: there are no independent reasons for the autonomy of world 2, but there are powerful reasons for this autonomy *provided* the autonomy of world 3 can be taken for granted. An argument which one might call an argument from human freedom and on which Popper seems to put great weight has exactly the same feature (*E*, 20ff; 206–55). Human freedom, Popper says, involves two things. It involves a certain *openness* (255) of the physical world, i.e. we must be physical indeterminists. And it involves '*means of control*' (239) in the domain thus released from the influence of physical laws: human actions are not always arbitrary, they are often the result of a deliberation (234) that is guided by ideas, plans, by a whole 'universe of abstract meanings' (230). This universe cannot be physical, or else we are back with a closed physical world without freedom. Here again we have a train of thought that shows autonomy of some processes (arguments, plans, ideas) *provided* the autonomy of some *other* processes (human freedom) can be taken for granted. Besides, the only remedy it offers for the 'nightmare' (222) of physical determinism is another, but 'autonomous' nightmare. For what is so great about a situation in which we are released from a physical prison to be imprisoned even more effectively by third-world walls such as the laws of logic, the laws of arithmetic, the laws of counterpoint?

animal kingdom' for world 3 (*E*, 23) and he points out that human effort 'may get a hold in a third world of critical discussion' (148) while animals are denied that privilege. This means either that there is no third world in the animal kingdom, or that there is something like a third world, but that it does not contain critical discussion. The last interpretation seems to be the correct one for we read, later in the book, that organisms, even amoebas, face 'objective problems' which 'need not have their conscious counterpart; and where they have their conscious counterpart, the conscious problem need not coincide with the objective problem' (242). Problems of this kind are created by the physical surroundings of the organism; but they also arise from the transformations which the organism effects in these surroundings and which have often unintended side effects; the transformations 'may create a new need, or a new set of aims' (117) and thus either act back upon the organism, or at least have the potentiality of so acting '[a]nd this potentiality or disposition may exist without ever being . . . realized' (116).[8] The interactions between an animal and his surroundings give rise to a whole 'universe of possibilities and potentialities . . .: [to] a world which ['is (both) . . . more abstract' than the world of physical bodies (116), and] to a large extent *autonomous*' (118, original italics).

With man the situation is exactly the same. Here, too, we have physical products, books, libraries, some of them created (written) by human writers, some manufactured in a purely mechanical way (for example, a book of random numbers, or of logarithms (115)), and these physical products are used in various ways. They are 'understood or interpreted, or misunderstood or misinterpreted' (116). They have the power, or the disposition, or the potentiality to be understood, or misunderstood, interpreted, or misinterpreted even if there is no one around who does the understanding and the interpreting. And again, all that is necessary for a book 'to belong to the third world of objective knowledge' is that it 'should – in principle, or virtually – be capable of being grasped . . . by somebody. But I do not admit more' (116).[9]

The thesis hiding in this passage may be formulated by saying that any *object* that possesses the power to make organisms act in a certain way is thereby shown to be an autonomous third-world object. This thesis makes world 1 and world 3 coincide: *every* object has the (either realized, or unrealized) power to make either man or some organism act in a certain way. Birds sit on stones, old shoes, trees, man reads the will of the gods in

[8] 'A wasp's nest is a wasp's nest even after it has been deserted' (115): apart from being a certain physical structure it also offers advantages, or resistances to properly equipped organisms, and it offers these advantages and these resistances even if there are no organisms around to profit from them, or to be disturbed by them.

[9] Popper continues: 'We can thus say that there is a kind of Platonic . . . third world . . .' which shows that possessing the *power* of being understood (or lived in, in the case of a nest, or sat upon, in the case of a stone) is indeed regarded as being sufficient for autonomy.

tea leaves, in the stars, in patterns in sand. Of course, this is not the only thesis we can find. Making full use of his favourite slogan that '*words do not matter*' (18, original italics),[10] Popper *also* asserts that it is the *powers* of the objects and not the objects themselves that form 'a . . . new universe' of autonomous entities.[11] He seems to argue that powers, being more abstract than the objects of which they are powers, belong to a different world.

6. ABSTRACTIONS CONSIDERED IN RELATION TO WORLD 1

This argument (or, rather, this *thesis*, for there is again no argument) shows that the criteria for autonomy of section 2 must be supplemented by rules that determine which objects do and which objects do not belong to the 'same world'. It is true, for example,[12] that the mass of a physical object cannot be reduced to its shape, and that laws about mass (such as mass conservation in reactions of a certain kind) cannot be reduced to geometrical laws. Mass is autonomous with respect to geometrical properties. But we should not on that account separate the mass of an object from its shape and transport it into a separate 'world'. For world 1 would then not be a world of actually existing material things, but an abstraction, and it would need the collaboration of various abstract worlds to arrive at something that can aspire to be 'the very standard of reality' (*E*, 21). The same applies to dispositions, powers, potentialities. Popper himself has suggested an interpretation of probability that makes it a real tendency (in world 1) of real physical objects. It follows that possession of a power is not sufficient for removing either the power or the object that possesses the power from world 1. And if it is pointed out that the argument is about *special* powers, viz. powers to evoke human (or animal) behaviour, then the reply is that such an argument again assumes what is to be shown, viz. that there are different kinds of powers, possessed by different kinds of entities, and forming different, and mutually autonomous worlds.

These considerations show that *abstractness* is not sufficient for ejecting an entity from the physical universe, as Popper seems to believe.[13] Energy, spin, parity are highly abstract entities, and yet they belong to world 1.[14]

[10] In most cases he uses this slogan to increase his own freedom to express himself. *Other* writers are treated much less leniently, especially when the words which don't seem to matter to them are his own (cf. the criticism of Sellars in *CR*).

[11] The two versions about who or what ascends to heaven number three occur only two pages apart. Page 116, line 19: 'a *book* should . . .'; page 118, line 1: 'a whole new universe of *possibilities*' (my italics).

[12] What follows is true of prerelativistic physics only. But *all* our arguments about autonomy refer to a certain stage of knowledge and may be overthrown by the arrival of a different stage.

[13] Meanings are 'something abstract' (*CR*, 298; cf. 116, 230, etc.) and 'abstractions do not belong to the physical world' (*CR*, 298).

[14] In a letter to me Imre Lakatos doubted that energy belongs to world 1. But energy has mass and what has mass belongs to the first world. Or does it not?

Physical laws are 'descriptions of the structural properties' – i.e. of something abstract, and yet 'of nature – of our [physical] world (196) as Popper says himself. What reason is there to assume that meanings, or numbers which are of a comparable, and perhaps even of a lower degree of abstraction (just remember that every conserved quantity is an invariant with respect to certain transformations), do *not* belong to the physical world? What reason is there to assume that works of art, music, theories are autonomous with respect to physical nature, and not part of it? This is the question that has now turned out to be important, and to *this* question Popper gives some most unsatisfactory answers. 'Logical relations such as consistency, do not belong to the physical world, they are abstractions', he writes (*CR*, 298). To start with, this is not an argument, it is a blunt assertion. Reputable people such as Aristotle or Hegel have asserted the opposite, and with excellent reasons.[15] Of course, the problem is a difficult one and cannot be settled in a page or two. But Popper tries to settle it in a line. Perhaps his argument is that logical relationships are autonomous *because* they are abstractions? This opinion has just been found to be unsatisfactory. It is to be admitted that we can always 'abstract' from the particular properties of a physical system and concentrate, say, on its energy, just as an economist 'abstracts' from the height, the weight, the intelligence, the sex appeal of people and considers their economical behaviour only. But this does not mean that the abstractions of the economist or of the physicist can in no way be regarded as aspects (properties) *of these bodies*. Quite the contrary: it is essential to remember that they *are* such aspects and that they *do* belong to the physical world, or the whole enterprise (e.g. economics, or the calculation of energy spectra) ceases to make sense. And there is no reason why this argument should stop short at logic. So my conclusion is the very opposite of Popper's. Abstractions, far from being excluded from world 1, must always be referred back to the physical situation from which they arose (or for which they have been designed) and of which they form an essential part. (In relativity energy has mass; and the same may be true of concepts.)[16]

Similar remarks apply to an objection which Popper raises against the naturalistic theory of art: '. . . the naturalistic theory', he intones, 'seems to me absurd. So far as the creation of music can be explained, it has to be explained at least partly in terms of the influence of other music . . . and, most important, in terms of the inner structure, the internal laws and restrictions . . . whose absorption and occasional defiance are immensely important for the musician's creativity' (*E*, 25) – undeniably correct; but

[15] Cf. Aristotle's arguments against the *chorismos* of the ideas.

[16] The question of what happens to abstractions *while they are treated as abstractions* will be dealt with presently. *Here* the problem is whether they can be regarded as parts of the physical world. They can.

where is the difficulty for the naturalist? Of course, if the naturalist restricts himself to the isolated description of noises (as Popper seems to assume: *E*, 25), then he will not get very far. But the physicist, far from resting content with an enumeration of single events, has long ago proceeded to a 'description of . . . structural properties' (196); so why should it not be possible for him to give a physical account of musical structures as well? Why should he not be able to describe the structural properties of man-made parts *of the physical world*; and thus provide the explanation 'in terms of other music' which Popper regards as so essential?

Both the argument from abstraction and the argument from the role of structures in the arts rests on a rather poor view of the physical universe. If powers, dispositions, invariants, structures are in principle excluded from world 1, then world 1 becomes a collection of relatively independent and unstructured events, rather similar to what Wittgenstein supposed it to be in his *Tractatus*. If only those powers are ejected which involve man in an essential way, then we want to know why man should be given such a special position. If the reply is that the powers in question influence the physical universe only via man, then we can point out that electric fields influence matter only via charge and yet nobody is prepared to exclude charge from the physical universe. Of course, there is a hidden assumption involved and it is that man is not entirely part of world 1. Are there arguments for this assumption? Yes, but they presuppose the existence of an autonomous third world, as we have seen. Turn as we may, the arguments we have examined do not give us a single independent reason for keeping abstractions out of the first world and for permitting some of them to influence this world only indirectly, via a third-world influence upon human minds.

7. ABSTRACTIONS CONSIDERED IN RELATION TO EACH OTHER; ARITHMETIC

So far we have looked at 'abstractions' in relation to the physical world and we have seen that nothing prevents us from counting them among its constituents. But abstractions are often examined in isolation and some of them, such as the ideas of pure mathematics or some dogmas of Christianity, seem to be not just *removed* from the physical world, but *irreconcilable* with it. Moreover, such entities seem to obey laws that are as independent of our mental processes as are the laws of physics, so that contemplating them we are led to discoveries very much like geographical discoveries in world 1 (74). Combining the apparently non-physical and non-mental character of abstraction *qua* abstractions with their independent behaviour, Popper arrives at some of his most impressive propaganda moves for world 3.

Popper admits that abstractions and even the natural numbers 'are the work of men, the product of human language and of human thought' (160). But having been introduced, they create side effects, difficulties that are 'in no sense made by us; rather, they are *discovered* by us; and in this sense they exist, undiscovered, before their discovery. Moreover, at least some of these unsolved problems may be insoluble' (160–1, original italics). Having introduced numbers

> we *discover* that the sequence of natural numbers consists of even numbers and of odd numbers and whatever we may think about it, no thought process can alter this fact of world 3. The sequence of natural numbers is a result of our learning to count – that is, it is an invention within the human language. And it has unalterable inner laws and restrictions or regularities which are the *unintended consequences* of the man-made sequence of natural numbers; that is, the unintended consequences of some product of the human mind [*E*, 22]

Two elements occur in this assertion. One is the objective character of the problems met and of the situations discovered. Both are 'real' in the sense of the reality criterion set up by Popper: they have effects upon the physical world. Occasionally this fact alone is supposed to establish their autonomy (e.g. 160, original italics: 'These problems are clearly *autonomous*. They are in no sense made by us'), but clearly *not* in the sense explained in section 2: the problems of a mountaineer, for example, are partly created by him (choice of the wrong path, of the wrong equipment); they have the same objective character as mathematical problems, and yet they can be explained in physical terms. Once more we find an argument that asserts autonomy but manages to establish reality only.

The second element of the assertion denies that numbers, meanings, and the problems to which they lead can be explained in physical terms and introduces autonomy in the full sense of the word. It is extremely difficult to see this second element in perspective and to discuss it properly. There exists such a strong predilection for non-physical numbers and meanings that any story in agreement with the predilection will at once be regarded as an additional argument in its favour. And the debate between realists and formalists (constructivists), empiricists and apriorists has been going on for such a long time and has led to such almost insurmountable complications that any opinion running counter to a familiar prejudice embedded in it will at once be ruled out of court as showing a sorry lack of sophistication. I cannot entirely get rid of the suspicion that it is this mixture of confusion and prejudice that has tempted Popper to venture into the philosophy of mathematics in the hope of gaining from it additional *and cheap* support for his beloved world 3. For the arguments he has to offer in this field are hardly impressive: autonomy of numbers is supposed to follow from the 'fact' (*E*, 22, second column, line 13 from below) that there are infinitely many of them, most of them neither thought about nor written down. Does this mean that atoms, or people, become members of world 3 as soon as there are infinitely many of them? Denial of an autonomy of meaning is supposed to rest on a 'trivial mistake' (298), i.e. on a confusion between two *senses* of 'thought' when the question is not how many senses of 'thought' *there are* but how many and which ones are *instantiated*. The arguments against a constructive approach boil down to the assertion that 'the autonomy of the third world is undeniable' (139) which in turn is linked to the 'fact', already mentioned, that there are infinitely many integers with complex and often unthought and undescribed relations between them. Popper asserts that this 'fact' is not incompatible with a constructivist approach (134) and that the appearance of incompatability arises only because of the constructivist's ignorance of unforeseen side effects which we *discover* rather than construct (138). Add this insight to constructivism and you see, according to Popper, that 'it can be combined with a kind of Platonism' (134). But where is the constructivist who denies discoveries? No constructivist is unaware of, or denies the existence of, unforeseen side effects of his constructions.[17] What he *does* deny is that the side effects include non-physical (non-mental) entities obeying non-physical (non-mental) laws and forming autonomous 'worlds' of their own. What he *does* deny is Popper's 'fact'. What has Popper to oppose to *this* denial? Only his own firm belief in the existence of the entities, a practice (naive quasi-intuitive arithmetic) which seems to support the belief *but for which some quite different interpretations have been offered*, as well as a rather naive pocket-history of the origin of numbers (*E*, 22). In addition, he neglects the results of more recent research in the foundations of mathematics which throw doubt on many of his assertions, if not on the idea of the third world itself.[18]

But, however bad his arguments, is he not at least right in his *conjecture* that numbers occupy a domain of their own and that a world-1 account of arithmetic is impossible? To decide *this* question, let us take a look at various layers of the practice of arithmetic, starting with the most primitive part, viz. counting.

8. COUNTING; SIMPLE ARITHMETICAL LAWS

Numbers arise from counting. The first stage, almost everywhere, seems to be the *construction of simple computers* (twigs, beads, series of marks on a wall, on a bone, series of memory images, something like an early version of the abacus, and so on) *and the discovery of various properties of these computers*. For example, one discovers the following law.

> If you have two heaps of apples, *R* and *S*, if you first make one mark for each apple of *R* on a wall, or on a bone and, adjacent to this series of marks, a mark

[17] The outcome of any consistency proof is unforeseen.

[18] I am here mainly thinking of the work of Imre Lakatos, starting with his paper in *Proc. Arist. Soc.* Suppl., 36 (1962).

> for each apple of *S* and then give names to the marks, one after another, the way you have learned it in the mark-series-naming schools, then you will end up with the same name whether you now start with *R* and continue with *S*, or start with *S* and continue with *R*.[19]

There is no doubt that this law is about the (physical) properties and dispositions of, as well as the relations between, the elements of three world-1 (or, if you wish, world-1 cum world-2) systems and processes, viz. (1) the apples; (2) the process of making marks and the marks themselves; (3) the naming procedure. Accordingly, it changes either when the apples and the marks change their nature, for example, when they start behaving like elementary particles, or when the procedure of correlating marks with apples is changed, or when the naming procedure is changed. Thus it is not true that 'no thought process can alter this [law]' (*E*, 22). If you want to count 'one, two, two, three . . .' instead of 'one, two, three . . .' every other time you count and if you carry out this wish of yours, then the 'law' will be 'altered' as a result. Similarly, involuntary malfunctioning of thought processes such as misremembering, or physical malfunctioning such as misplacement of signs will affect the law.

Now once it has been discovered that the law is valid only for certain methods of counting and correlating marks with things,[20] and once one has realized its usefulness for dealing with large assemblages (if such assemblages are 'counted' in accordance with methods that guarantee the validity of the law, then the result of any counting procedure will be independent of the order of counting), one will of course want to ensure its continued usefulness by ensuring that everyone sticks to the right method. One will *train* mathematicians and even common citizens in such a way that the law (along with other laws that are discovered as the practice of arithmetic expands) is preserved. I do not see how this training can change the 'ontological status' (161) of either the law or of the processes that conform to it. (Do the parts of a cash register change their ontological status because they have been put together with a certain purpose in mind? And do statements about the way in which these parts behave cease to be empirical statements just because the parts have been shaped by man with a purpose in mind?) Nor does the training make the laws unalterable. We may still decide to count differently and in this way change the law (manufacturers of cash registers may decide to build registers that favour the grocer). Or we

[19] Note that the correlations between marks and objects do not involve classes, or classes of classes. We simply correlate two perceptually discontinuous structures, 'parts' of the one being attached to 'parts' of the other (what is a 'part' is here decided by perception alone).

[20] I do not think that any such 'discovery' ever took place and played a role in the history of mathematics. It is much more likely that law and methods of counting entered a kind of equilibrium state without much deliberation on the part of the counters and that the equilibrium state was preserved in schools. However this does not affect the result of my argument which depends on the *components* of the equilibrium state and not on its *origin*.

may err despite all the training we have received (the cash register may malfunction despite all the care spent on its construction).

Of course, the notion of an error presupposes a standard of comparison, but does this standard have to be non-physical? Animals and human parents correct the errors of their young by physical actions which emerge either from the 'innate structure (the "programme") of the organism' (72) or from the results of a learning procedure similar to the one to be carried out. Large factories and undertakings such as the moonshots are 'supervised' by computers which correct deviations from the standards (from the basic programme) by sending physical signals to the malfunctioning parts thus correcting the malfunction.[21] In the case of simple arithmetic 'laws' such as the one we are discussing here the 'standards' are simple computers such as the multiplication table combined with simple rules for counting and inserting results into specific computations. (The standards need not be written down, or exhibited so that they can be *seen* by all; they may be part of an *oral tradition* that is sufficiently stable and accessible to function as a point of comparison, and that simplifies counting, multiplying, etc. in the way indicated in the law.)

Now once the standards or, in physicalistic language, the master computer has been chosen, we have '*objective*' ways for evaluating a particular move such as a particular calculation; we simply connect it with the master computer to check the outcome. Proceeding in this way, we can say 'you have made a mistake, try again' *not* because numbers (at the present stage of simple arithmetic) are non-physical and non-mental entities, but because part of the practice of arithmetic (the counting tradition; the standard sequence of thoughts evoked by the statement of an arithmetical problem; the master computer which may consist of relatively stable memory images properly arranged and available on short notice) is sufficiently stable, sufficiently accessible, sufficiently independent of the more fleeting processes in our minds, sufficiently easy to use, to provide a point of comparison for more casual events and because we have arrived at a stage of knowledge where we let this particular stable part rule the rest.[22]

[21] Let no one object that the astronauts provide the 'human element' and thus introduce world 3. Whoever has watched them in their space capsule or has heard them talk on earth will realize that they functioned as computers during the flight (readings of Genesis not excluded) and are incapable of functioning differently even when returned to their private lives. Cf. Norman Mailer, *Of A Fire On The Moon* (New York, 1969), Ch. 2, 'The Psychology of Astronauts'.

[22] It is the stabilizing aspect of the third world that has impressed Lakatos: the outcome of a calculation, of an experiment, of an argument depends on 'objective' criteria, it does not depend on the whims of the calculator, or of the person arguing. This, of course, is not true. Nobody can be forced to accept any result, however 'rational', and anyone can escape the force of the most stringest argument by simply saying 'I don't like it'. On the other hand, he may decide to submit to the rules of one of the many computers he is carrying in the memory part of his brain, or in some external physical system. If the computer is built in a stable way and contains easily identifiable parts, then it will be easy to apply the rules and to judge

Standards so conceived are *not infallible* in the sense that they invariably satisfy the purpose for which they were introduced (in our case, to make the result of counting independent of the order of counting), and they must occasionally be changed.[23] All that is asserted is their objectivity. *And as we have seen, this objectivity can be guaranteed without any ascent to a higher world.* The laws of arithmetic which we have encountered so far can be objective without being autonomous (in the sense of section 2).[24]

9. INTERLUDE: OBJECTIVIST ACCOUNTS OF KNOWLEDGE

One of the main motives for Popper's third world is his dissatisfaction with what he thinks is the rampant subjectivism of contemporary epistemology. Now I think he has overrated some of this subjectivism. When contemporary epistemologists try to explain knowledge as well-supported belief, they don't mean any kind of belief, however fleeting and incomprehensible; they mean belief that has been standardized in a certain way and has become part of a tradition. Popper is of course right in pointing out that a tradition is never restricted to what goes on in the minds of the participants, but finds expression in a language, in scrolls, books, drawings, diagrams, decorations, works of art. A theory of knowledge must examine all these objects, it cannot rest content with a study of belief alone. Popper's suggestion leads to a much needed enrichment of the subject. It is not new, as we shall soon see. As a matter of fact, theories of knowledge of the type envisaged by Popper were put forth (to choose a few examples at random) by Aristotle, by the Church fathers and, in our century, by Wittgenstein. What *is* new in the case of Popper is his attempt to extend the Bolzano–Frege idea of a third world to the entire field of culture and to admit change into it. I don't think the attempt is successful, or likely to advance our understanding, and I

cases, with their help, in an 'objective' manner. We may continue using the phrase 'world 3' as a *figure of speech*, describing the relatively stable, unalterable and, because of their connection with tradition, intersubjective parts of our thought processes as used in conjunction with other types of computers (formalisms on paper, languages, etc.). All the arguments of Lakatos which involve a 'third world' remain valid in this interpretation. And he can now distinguish between two types of change in a research programme: changes that affect the master computer, and more peripheral changes that affect auxiliary computers only and are usually pushed aside as 'errors'.

[23] This happened to certain standard ways of relating elements to classes after the discovery of the logical paradoxes.

[24] One of the strongest reasons for making an ascent to a third world is the loaded terminology that is used for describing certain parts of the practice of arithmetic. One speaks of 'errors', and errors are not just differences, but differences *evaluated in a certain way.* But mention values, and everyone is ready to leave the physical world that is being evaluated. Well, it is easy to provide a physical model for that aspect of the matter as well. Consider a machine capable of pattern recognition that rejects patterns differing from the standard by more than a certain amount, and you have all you need for a physicalistic account of the *practice* of error rejection. Of course, a very common *ideology* of error rejection is still unaccounted for and rightly so, for the practice can get on very well without it.

shall soon produce more arguments to that effect. To see it in perspective, let me now describe a few objectivist accounts which satisfy Popper's demands not to rest content with belief alone.

According to Aristotle knowledge is neither an unchangeable Platonic heaven, nor a special state of the soul. It is a complex social product that influences every generation, may even dominate it, but which is also changed and augmented by every generation:

> The investigation of the truth is in one way hard, in another easy. An indication of this is found in the fact that no one is able to attain the truth adequately while, on the other hand, we do not collectively fail, but everyone says something true about the nature of things and while individually we contribute little or nothing to the truth, by the union of all a considerable amount is amassed. [*Metaphysics*, 993a30ff]

Occasionally an author is lucky and can lay the foundation for an entirely new subject. 'I set myself the task', says Aristotle in the concluding remarks to his *Topics* (183a37ff)

> to find a method which would enable us to draw conclusions concerning a proposed statement starting with generally acknowledged views. This is the task especially of dialectics, but also of scientific inquiry . . . It is clear that I have reached my aim. However,

Aristotle continues

> I must remind the reader of certain things concerning this lecture. Concerning everything that has been found or invented the situation is such that what has been achieved by the predecessors and taken over by their followers has later on been gradually augmented.

Aristotle's belief that knowledge is an objective and ever-changing social product, and that it can be improved only by the collaboration of many generations was responsible for his grandiose attempt to collect opinions from all fields and to reconstruct their development, starting with religion and myth, and gradually proceeding to his contemporaries:[25]

> Our forefathers in the most remote ages have handed down to posterity a tradition in the form of a myth, viz. that the [celestial] bodies are gods and that the divine encloses the whole of nature. The rest of the tradition has been added later in mythical form with a view to the persuasion of the multitude . . . They say that these gods are in the form of men or like some of the other animals, and they say other things connected with and similar to these which we have mentioned. But if one were to separate the first point from the additions and take it alone – that they thought the first substances to be gods – one must regard this as an inspired utterance and reflect that, while probably each art and each science has been developed as far as possible and has again perished, these opinions with others have been preserved like relics of the ancient treasure. [*Metaphysics*, 1074blff]

[25] For Aristotle's rather enlightened attitude towards myth, cf. Werner Jaeger, *Aristotle* (Oxford, 1958), ch. 6.

Aristotle thus became the first and, perhaps, the most outstanding historian of ideas in Western thought.[26]

His view of knowledge also influenced the shape of his *arguments*. Every discussion, every attempt to arrive at an acceptable solution for a difficult problem starts with a 'logical reconstruction' of older views; we hear how these views have changed and in what respect they are still unsatisfactory until we finally arrive at a new solution that preserves what is valuable and eliminates error:

> For our study of the soul it is necessary, while formulating the problems of which in our further advance we are to find the solutions, to call into council the views of those of our predecessors who have declared any opinion on this subject, in order that we may profit by whatever is sound in their suggestions and avoid their errors. [*On the Soul*, 403b20ff; note the emphasis on *problems* which plays such a large role in Popper's philosophy.]

We see that Aristotle's theory of knowledge contains most of the elements Popper wants to emphasize. It has the right amount of generality and sophistication, not too much, not too little. And it has borne fruit. It has led to new subjects and new ways of solving problems. To my mind it is all we need for an answer to the subjectivists of our day.

The view that knowledge is an objective product that is largely independent of our interests and our wishes and is handed on from one generation to the next, plays a most important role in the history of Church dogma. Reacting to various heresies that were based on a certain reading of the Scriptures, the Church fathers, and especially St Irenaeus and Tertullian, recognized that a collection even of inspired writings does not suffice and that we need canons of interpretation. The canons cannot be set up in an arbitrary manner, but must have some authority. For St Irenaeus the authority is historical: we accept that reading and that general philosophy that can be traced back to the apostles via an unbroken chain of generations, starting with the first bishop of Rome:

> The tradition of the apostles is manifest in the whole world; it needs only to be examined in any church by one who wishes to know the truth. We can number the bishops who were instituted by the apostles and their successors down to ourselves . . . We can show that the tradition which [the Church] received from the apostles and the faith it has preached to men have come down to us through a succession of bishops. [*Adv. Haer.*, III, iii, 1–2]

> And if some little question should lead to a discussion, should we not have recourse to the oldest churches, those in which the apostles lived, and in the matter under discussion receive from them a certain and manifest doctrine? And if the apostles had not left us the scriptures, would it not have been

[26] Greek views concerning the development of objective knowledge are discussed by Ludwig Edelstein in *The Idea of Progress in Classical Antiquity* (Baltimore 1967). According to Edelstein, the first philosopher to discuss the properties of objective *human* knowledge was Xenophanes.

necessary to follow the order of the tradition they had entrusted to those to whom they confided the churches? [*Adv. Haer.*, III, iv, 1]

(Luther attacks this traditional element of the Church of Rome which was constantly augmented by papal encyclicals, by the decisions of councils, by philosophical speculation, and replaces it by the subjective understanding of the individual reader. This, and the emphasis on individual conscience was one of the roads to subjectivism.)

And of course there is Wittgenstein who emphasizes the need for objectivity, and who therefore examines the material manifestations of knowledge rather than beliefs, sensations, feelings and the like.[27]

All these theories, assume, or explain, objectivity without invoking autonomy. Moreover, Wittgenstein has gone much further in his account of mathematics (including arithmetic) than has Popper. He has detailed answers to questions which in Popper are solved by an appeal to autonomy.

I now continue with the discussion of arithmetic.

10. ARITHMETIC (CONTINUED); HOW TO TURN AN OVERSIGHT INTO AN EXISTENCE PROOF

The 'law' discussed in section 8 is a rather simple law and it belongs to a rather simple stage in the development of arithmetic. We have seen that it can be stated and that its features can be discussed without invoking 'third world' objects'.[28]

Next comes a stage where laws are stated independently of the process (of counting) which they characterize. Let '*m*' be the name of the last mark associated with the elements of *R*, '*n*' the name of the last mark associated with the elements of *S*, '*s*' the name of the last mark of the combined series *nm*, '*r*' the name of the last mark of the combined series, assuming we do our counting in the opposite direction, taking first *R*, and then *S*, then our law can be expressed by saying that *r* and *s* are the same name, or that LAST MARK (*mn*) equals LAST MARK (*nm*) or, further abbreviated, *mn* equals *nm*. We may also generalize and assume the result to be valid for *any* pair of heaps, however large, and consisting of *any* constituents.[29]

[27] Cf. his *Philosophical Investigations*.

[28] The objection that we are not simply dealing with processes but with dispositions as well has been answered in section 6: dispositions are part of the objects whose dispositions they are.

[29] The generalization is *false*, as can be seen when considering assemblages of amoebae, droplets of water, or states of elementary particle systems. We can make it true by *ad hoc* adaptations, such as the assumption that the objects counted remain unchanged during the process of counting *as regards their number*, or by changes in our counting procedures. A *general* injunction against *all* change would make the law inapplicable: everything changes in some respect all the time. 'Pure' mathematics is empirical mathematics protected by a complicated system of *ad hoc* moves. Cf. I. Lakatos, *Proofs and Refutations*, (Cambridge, 1978). Cf. also n.31 and section 11.

The 'law' we obtain in this way may be written

$$mn = nm$$

and it may be said to have two 'sides'. On the one side, there is the *regularity* it describes; on the other side, there is the *formula* used in the description together with the *statement* expressed by it. I shall now discuss these 'sides', in order.

The *regularity* belongs firmly to world 1 (and world 2). It is a regularity exhibited by complex systems, comprising physical and mental parts, the latter including thoughts and thought patterns of considerable rigidity, as well as dispositions for producing such thought patterns. It belongs to the psycho-physical world, as we have seen in section 6, when talking about the 'ontological status' of abstractions, and in section 8 when we discussed a more complete formulation of the law.

This result may be stated by saying that *the law*, which for most modern thinkers belongs to a special discipline, arithmetic, *can be reduced to physics*. Of course, there exists no set of physical conditions *outside* the counting organism which, taken together with known physical regularities, will give us this law. A law that is sustained by the regular behaviour of some organism or some computer can be 'reduced' only if the structure of the organism (its programme) or of the computer is included in the initial conditions. A law that reports the regularities one finds *when counting* can be reduced only if the peculiarities of the counting procedure are included in the initial conditions. Some older empiricists overlooked this requirement, and they failed.[30] Popper also overlooks the requirement but, being somewhat less hesitant about a multiplication of entities, he can turn the *failure* into the *discovery* of a third world. One easily sees how this comes about. If we omit the counting brain and its dispositions from the reduction, then '$mn = nm$' can no longer be reduced to physics. It becomes an irreducible formula. Irreducibility, for Popper, means autonomy: we have discovered some autonomous entities. What entities? Why, the entities the law describes! Now, strictly speaking the law no longer describes any entities. It was introduced to express complex relations between heaps of objects, marks, and an organism that behaves with a certain regularity. The organism has been omitted, no proviso has been made to reconnect the law with reality,[31] so its content remains unspecified. This certainly is a disadvantage, but not for a determined propagandist. For now its content can be

[30] Mentalists, on the other hand, disregard the physical part of the law, the part played by individual marks, walls, bones, etc. and their reductions are incomplete in another respect. Here again some critics took the incompleteness as a sign of otherworldliness. One example is Husserl. Another is Frege.

[31] This is sometimes explicitly admitted by those who speak of being interested in 'pure mathematics' only. So the mess receives a name, but it is not removed.

changed at will. It can be changed at will without there being the impression that any change has taken place (after all, we have been conditioned to believe that the law has something to do with numbers). So Popper, having 'proved' autonomy, can at once say that the law is about numbers, and that numbers are autonomous non-physical entities. And he even receives praise for his acumen, for the non-physical interpretation of numbers is quite popular.[32] He receives praise because he has not only accepted the 'correct' interpretation, but also shown it to be true. Next, he 'discovers' that numbers have an effect on the real physical world. There cannot be a *direct* effect (cf. the argument in section 3), and we need a mediator. The mediator is soon found to be that marvellously mysterious entity, the human mind (*not* the brain – *that* would be too concrete and too specific!). Having first been omitted, the mind now returns to the scene. But it does not return in the shape in which it left. Interposed between a physical world and an 'autonomous' mirror image of the original omission, it looks rather strange, which proves that it belongs to yet another world, world 2. And so discovery follows discovery until the entire structure finally stands in front of us in its full splendour: an 'open' physical world (*E*, 26), its 'gaps' filled with world-2 actions guided by the contemplation of world-3 objects. How thrilling, and how absolutely original! But the 'openness' is of course all of Popper's own making. Having omitted the counting organism from the argument, he notices a gap; being favourably disposed to gaps in the physical universe he interprets it as a sign of autonomy, and fills it with non-material entities ('Indeterminism is not enough'!, 226). We have an oversight, and a tendentious interpretation of its effects, that is all. Some of Popper's most impressive arguments for emergence and autonomy (in the strong sense of section 2, not in the weak sense on which he frequently falls back, just to be on the safe side; cf. section 4) work only because of this oversight, and they collapse with its discovery.[33]

Now we have seen that some of Popper's arguments for the third world, such as the argument from abstraction, involve a rather poor view of the physical universe: if abstractions such as powers, dispositions, invariants, and structures are excluded from world 1, then world 1 becomes a collection of relatively independent and unstructured events, rather different from the rich and complex physical world that science has built up around

[32] For the source of this interpretation and some arguments against it, cf. the next section.

[33] As an example, cf. Popper's argument for the irreducibility of the name relation as reported in n.6: In the world *outside the observer*, causal chains have no beginning and no end; the name relation has a beginning and an end; hence, the name relation cannot be reduced to physics. But the reduction is quite straightforward once we include the pattern-recognition mechanisms of the observer in the premises. These mechanisms are purely physical and some of them start working already in the retina. (Employing Popper's argument in the presence of sunglasses, we could prove the existence of irreducible shadows darkening the entire physical world.)

us and that impresses Popper so much on other occasions. The present discussion makes us suspect that Popper has a poor view of the human mind as well: if an organism behaves in a certain regular way in a world that, taken by itself, does not seem to possess such regularities, then the reason is often that he perceives *another world*, i.e. he *reacts passively* towards non-physical entities (such as the famous unforeseen consequences of the past activity of his species, or his tribe), rather than *proceeding actively* in accordance with some slowly developing psycho-physical programme.[34] Popper *does* consider the second alternative, but all he has to say about it is that it is 'evasive' and provides an 'easy way out' of 'this difficulty' (225). Now, first of all, what is wrong with an 'easy way'? Should one heap complexities upon complexities when a few words can do the trick? And, secondly, what is 'this difficulty' one is supposed to solve in a difficult way? 'This difficulty' is the question of how abstract non-physical entities can affect the physical universe. It arises only if one postulates such entities in the first place. We have seen that the arguments which Popper has to offer on their behalf are anything but convincing and that his most impressive speeches rest on an oversight. True, there is the general *impression* that numbers are something out of this world, and there is also a *theory* which implies that the impression is essentially correct: *there are* non-physical entities with complex relations between them. Now which procedure is more 'easy' and 'evasive', a procedure that just *accepts* the impression, *repeats* the theory, strengthening it here and there with sham arguments and heavy rhetoric (cf. the remainder of p. 225), or a procedure that tries to show how the impression of otherworldliness arises from an incomplete account of the interaction between man and the universe, that opens up new avenues of research (cf. the paper by Hubel and Wiesel, quoted by Popper on p. 72, which shows that some classifications occur already in the retina, and before the incoming information has reached the brain) and thus encourages content-increasing developments?

With this in mind, let us now ask the question: what is the 'ontological status' (161) of the *formula* that expresses the law? The answer is obvious. The formula belongs to world 1. If one objects that the interesting thing about the formula is the type, not the token, then the reply is either that types which are shapes, patterns, etc. are just as much part of world 1 as the material that has those shapes (see again section 6 on abstractions), or that the classification into types is carried out by the (either innate, or manufac-

[34] The word 'programme' is here used to describe the known and unknown, innate and manufactured structures of the mind that are responsible both for classifications that mirror the physical world and for other classifications that do not (false theories, works of art, dreams). The programmes change as a result of the interaction between man and the physical world.

tured) 'structure (the "programme") of the organism' (72), where the manufacturing is done either by teaching, or by less obvious methods of conditioning, or by both. In the latter case types are the result of an interaction between two physical systems, and they belong to the same 'world' as either of them.[35]

The *content* of the formula, or the statement expressed by it can be explained in a similar way. Taken by itself, and removed from the counting procedures, the formula does not express anything and it is futile to expect that we shall find its 'meaning' by gazing at some third world.[36] Connected with the standard counting procedures, it expresses the law I have formulated at the beginning of section 8. It expresses it *while being part of the process that constitutes the law*. Of course, the formula can be written down independently of any specific act of counting, just as one of the numerals in a cash register can be detached from it and exhibited all by itself. But just as the 'meaning' of such a numeral becomes clear when we see how it functions in the process of exchanging money and goods, in the same way the 'meaning' of the formula becomes clear only when we see how *it* functions in connection with the process of counting. All we need in order to grasp this function is a certain ability for pattern recognition. There is no reason why it should not be possible to give a physicalistic account of such an ability and of its effects, when exercised.

Now instead of counting simple physical objects we may decide to count vacant positions in a business enterprise, droplets of water, thoughts, hostile exchanges between distant countries, and so on. The law will remain valid on some occasions; it will not remain valid on others. Realizing this we may decide to count only those things that satisfy the law when counted in the usual way, and, taking the decision for granted, we may in future omit any reference to the counting procedure. Seeing the great variety of things counted and the great variety of procedures used in counting them, we may even feel inclined to abstract from *all* practices and hope in this way to arrive at the 'pure' (in the sense of 'pure mathematics') meaning of '$mn = nm$'. The hope is as illusory as is the hope of Kant's famous bird who, considering that flying in clean air is easier than flying in smoggy air, concluded that flying in no air would be best of all: if we abstract from every practice, then all we have are the formula and the pious hope that '*the* content' will now reveal itself. But we are luckier than the poor bird. The gaps we have created can always be filled with monsters of our imagination. This brings me to the last item in my discussion of simple arithmetical laws.

[35] For an ingenious anticipation of such an account, cf. Plato. *Theaetetus*, 156cff.

[36] Unless we give a formula a new kind of meaning by connecting it with a *metaphysics* of numbers. Cf. the next section.

11. THE METAPHYSICS OF NUMBERS

So far our imaginary history of arithmetic has proceeded to a stage where formulae such as $mn = nm$ are used for expressing complex psycho-physical laws, where more 'refined' practices such as counting in one's head are used side by side with more crude computers, and where one *might* feel inclined to disregard all practice and contemplate pure meanings. It is also assumed that for every collection, however large, marks and names for marks can be provided. To facilitate the process, names may be built up in a standard manner which leads to the decimal system and other notational systems. But now there comes what Weyl has called the 'leap into the beyond' of abstract entities such as numbers and classes, and totalites of such entities. This 'leap' involves two elements: (1) 'numbers' are correlated with number signs and are separated from the practices in which the signs initially received their meanings; (2) statements involving totalities of such 'numbers' are made. I shall discuss the first step only.[37]

This step seems to have been made in some mythical traditions where numbers played a special role and then, in a slightly modified way, in the school of Pythagoras. We have here a fascinating *cosmological hypothesis* that goes far beyond anything done with numerals up to the time of its invention, and that replaces the content lost by the removal of the practices.[38] The hypothesis is strengthened by Parmenides's theory of The One, which leads to a new, non-material interpretation of arithmetic.[39] The mental and physical computers which are still used as calculating aids – after all, numbers do not add and subtract all by themselves – now vastly increase in complexity, precision, rigidity. Moving in their own way, without any apparent interference from the will, from passing moods, or indispositions, they create the illusion of being guided by equally precise agencies (certainly not by apples!) when all they do is follow regularities that have been impressed upon them in the past. There was of course some guidance when the process of impressing first occurred, and this was due to a complex psycho-physical interaction, as we have seen. But the new theory of numbers denies such crude influences. The guiding agency must be sought in a different, non-material world.[40] It is sought, *and soon found*: just as the ancient believer, prepared by his ideology, virtually *perceived* gods, demons,

[37] For the second step the reader is referred to the literature on constructivism. Popper adds no new arguments to this aspect of the issue.

[38] In actual history the steps, of course, did not follow each other in such an orderly way. Whatever the sequence, the arguments concerning each step and any combination of them remain the same.

[39] Cf. the story as told in A. Szabó's fascinating book, *Anfänge der Griechischen Mathematik* (Budapest, 1969).

[40] Originally, this world was interpreted astronomically, and numbers were constellations. Cf. R. Eisler, *Weltenmantel und Himmelszelt* (Munich, 1900).

wood sprites – the regularities impressed upon his thoughts forming an essential part of his perceptions – in the very same way the number mystic and his modern follower now perceive non-material numbers and their relations.[41] *But the theories that engender such perceptions have been refuted and abandoned long ago*.[42] We therefore need alternative accounts of the impression of otherworldliness which we find in the domain of numbers.

12. CONCLUDING APPRAISAL OF THE THREE WORLDS

Popper's theory that admits change into the third world may be regarded as one such alternative account. It is badly argued, frequently *ad hoc*,[43] and the Platonism it still retains is nothing but the result of a simple oversight. The account I have sketched is another alternative. Far from merely admitting change into the Platonic heaven, it removes this heaven as being based on a simple mistake. But it retains all those features of our arithmetical practice which support its objectivity and which were wrongly thought to support its autonomy as well. Moreover, it makes new suggestions, it has led to interesting discoveries in psychology, neurophysiology, history, mathematics (constructivism), and it is bound to progress rapidly once all these discoveries are combined and subordinated to a single research programme. The balance *could* be changed in Popper's favour, if there were some knock-down argument to support his case or some discovery to show its fruitfulness. But there is no such argument, and no such discovery. That decides the matter: the third world is nothing but a chimera, a shadow cast upon our material world by views which no one in his right mind would now defend, and lengthened beyond recognition by rhetoric and insufficient analysis.

Now we have seen that the only argument in favour of an autonomous *second* world which Popper has to offer involves the third world in an essential way. If there is no third world, then there is no second world either. Thus materialism (or physicalism), freed from the blemishes which obscurantist 'arguments' seemed to perceive in it, once more emerges as the most promising research programme in existence.

[41] 'We operate with these objects [numbers, etc.] almost as if they were physical objects' (163).

[42] Cf. Popper's criticism of Parmenides as discussed in the paper referred to in n.3.

[43] Popper applies his position to various problems such as the problems of 'hermeneutics'. The applications involve a kind of 'third world', but one that can easily be explained in first-world terms. Moreover, his opponents, e.g. in the case of Galileo, are not people who deny the third world, but people who have strange views about methodology, who believe, e.g. that theories inconsistent with accepted basic statements should be given up. Some Popperians, such as Professor Hans Albert of Mannheim University, have tried, long ago, to supplement Popperian methodology with psychological hypotheses and thus to make it more realistic. Unfortunately the boss never gave their ideas the attention they deserve.

13. CRITICAL RATIONALISM AND ITS ANCESTRY

This concludes my review of the ontological parts of the book and especially of the idea of an autonomous third world. I shall not comment on the more specifically biological suggestions Popper has to make except to remark that evolutionary ideas were introduced into epistemology by Spencer, Mach, and Boltzmann and that these thinkers went much further than Popper: if knowledge is part of evolution, then even its most basic constituents such as the laws of logic, or the laws of arithmetic, or the use of argument as an instrument of progress must be regarded as temporary phases which are in need of improvement and which will be overcome at a later stage[44] (some of us believe that this 'later stage' has already arrived). Turning now to the passages which deal with methodology we are especially interested in those articles where Popper promises a 'full answer to [his] critics' (2). Popper, after all, claims to have invented a new, non-justificationist epistemology that has created a lot of interest and a lot of opposition, and one wants to know what has happened since it first appeared in book form in 1934. One would also like to know something about its predecessors in the nineteenth century, before the Vienna Circle and Popper appeared on the scene, for the rise of Darwinism and of the liberal theories of the time was likely to encourage and, as a matter of fact, *did* encourage non-justificationist methodologies even then. It is not easy to find this information either in Popper or in the writings of his followers. Popper introduced his view when the non-justificationist attempts of the nineteenth century were largely forgotten and when the idea of a science without justification and of a knowledge without foundations seemed not only new but also utterly absurd. This explains some of the historical myopia found within the Popperian school. (In the case of Popper himself the myopia can assume astounding proportions.)[45] To correct it, let us start

[44] Cf. Ludwig Boltzmann, *Populäre Schriften* (Leipzig, 1905), 318ff, 398f and *passim*. Cf. also my article on Boltzmann in *The Encyclopedia of Philosophy* ed. Paul Edwards (New York, London, 1967).

[45] Two examples. First, on p. 8 Popper writes 'Neither Hume nor any other writer on the subject before me has to my knowledge moved on from here [impossibility of justifying reasoning from experienced to unexperienced instances] to the *further questions*: Can we take the "experienced instances" for granted? And are they really prior to the theories?' Reply: the need for a theoretical evaluation of observations was realized by Aristotle. Key passages: *On the Heavens*, 306a18, *Prior Analytics*, 46a20 (only some data are reliable, others are not); *Metaphysics*, 1010b14 (scale of reliability, based on theoretical considerations); even sensations which admit hardly any falsehood (*On the Soul*, 428b18) may go awry because of mistakes 'in the operations of nature' similar to 'monstrosities' (*Physics* 199a33: sensations, after all, obey the general laws of motion); *On the Part of Animals*, 644b25 (dependence of reliability on the position of the object); *On the Soul*, 425b23 (influence of the structure of the sense organ; cf. *On Dreams*, 460b23 on the origin of illusionary perceptions); *On Divination by Dreams*, 462b14 (why false opinions are sometimes supported by experience); *Meteorology*, 355b20 (imperceptible occurrences); and so on. In Newton 'phenomena' can be, and often

with a short sketch of the development of methodology from Newton to Lakatos.

After Aristotle, the philosophy of the empirical sciences began with the attempt to explain in what sense general theories could be *justified*, or *proved*. Hume showed that justification can be neither derivation from experience nor probabilification. His arguments had hardly any effect on the sciences. They were unknown. Besides, they did not use the language of mathematics. Using the language of mathematics Newton had apparently succeeded in deriving the law of gravitation from phenomena. This 'derivation' dominated the scene right into the twentieth century.[46]

The mistakes underlying Newton's derivation were explained twice, first by Hegel, then by Duhem. According to Hegel,[47] the formula expressing the law of gravitation can be interpreted in two different ways. One interpretation makes it simply a transformation of Kepler's laws. The second interpretation asserts the law of gravitation. It cannot be derived from Kepler's laws. The impression of success arises because the writer and the reader change interpretations in the middle of the derivation.

Duhem showed that the law of gravitation contradicts Kepler's laws and cannot therefore be derived from them. Duhem was much easier to understand then Hegel. His argument received psychological support when Einstein gave an example of two different theories 'justified' by the same body of evidence. Hume's arguments re-entered the scene at about the same time. Once more it became clear that the problem of justification was still unsolved.

Among the older solutions the following three are worth mentioning: Kant's transcendental idealism; conventionalism; probabilism.

Probabilism admits that theories cannot be derived from facts. However, it is possible to determine their probability and to evaluate them on that basis. Lakatos has shown, in a long case study, why the probabilistic solution is unsatisfactory and how it became so.[48]

are, corrected 'from above' (cf. ch. 2). Mill says in *On Liberty* (quoted from Marshall Cohen (ed.), *The Philosophy of John Stuart Mill* (New York, 1961), 208) that experience alone cannot decide about the fate of a theory: 'There must be discussion to show how experience is to be interpreted' and discussion involves alternative views. Boltzmann quotes Goethe's well-known dictum that experience is only half experience and uses it in a criticism of what we today would call basic statements (*Populäre Schriften*, 222 and *passim*). And so on.

The second example occurs on p. 205. Here Popper criticizes some writers for referring to my papers rather than to his when discussing the inconsistency between higher and lower level laws, and points out that I myself quote Popper as my source. He makes two mistakes. First, he treats my reference to him as historical evidence when in actual fact it was intended as a friendly gesture. Secondly, the inconsistency was of course discussed before, by Hegel and then by Duhem, as Popper says himself (358).

[46] Cf. Max Born, *Natural Philosophy of Cause and Chance*, (Cambridge, 1949), 129ff.

[47] *Encyclopädie der Philosophischen Wissenschaften*, ed. G. Lasson (Leipzig, 1920), par. 270.

[48] Imre Lakatos, 'Changes in the Problem of Inductive Logic' in *The Problem of Inductive Logic*, ed. I. Lakatos (Amsterdam, 1968).

Conventionalism says that a theory is justified because it brings some order into the known facts and provides concepts and ordering principles for things as yet to be discovered. The order is never complete, for there are always recalcitrant phenomena. This does not invalidate the chosen scheme but challenges its defenders to rebuild the phenomena until they fit into it. The scheme is chosen either because it easily accounts for some empirical regularities (empirically motivated conventionalism), or because it follows from certain theoretical postulates (the Dinglerian version). Once the scheme is accepted and incorporated into scientific practice, it quite automatically transforms the facts and thus removes any criticism 'on the basis of experience'. Transcendental idealism assumes that man classifies data automatically, without having made a conscious decision to adopt an ordering scheme. The laws that underlie his classifications are found, and proved, by transcendental analysis.

Neither transcendental idealism nor conventionalism (of the empirically motivated kind) are universal theories. They offer means of justification for some very general views, but they become inductivist when dealing with simple regularities. And even in the case of the preferred views, the solution depends on empirical laws: Kant assumes that the human mind does not change, the conventionalists assume that the decision of the scientist to stick to certain forms of thought will keep these forms of thought imprinted on his material forever. The problem of justification moves from one place to another; it is not solved.

It remains unsolved in yet another theory that was formulated twice in the nineteenth century, first by Mill, in his immortal essay *On Liberty*,[49] and then by thinkers such as Boltzmann who extended Darwin to the running of science itself. This theory takes the bull by the horns: theories *cannot* be justified, and they *need not* be justified. The procedure of science is not to praise the good, but to eliminate what is bad, or comparatively inferior. The 'fundamental difference between [this] approach and [inductivism, or justificationism] is that [it] lays stress on *negative arguments*, such as negative

[49] It was John Watkins who long ago drew my attention to Mill and made me realize his importance in the history of critical rationalism.
Added 1980. Mill is often represented as an inductivist; and this he is, in his *Logic*. However, he has formulated his inductivism in a way that greatly facilitated the transition to a purely hypothetical account of scientific knowledge. Mill explains how hypotheses are introduced, how their consequences are developed and how they are tested. This part of his account is indistinguishable from Duhem's and, therefore, from Popper's. Mill regards hypotheses as legitimate parts of science and he praises some contemporary scientists (e.g. Darwin) for their interesting hypotheses. But he adds that some hypotheses can be proved by showing that no others can account for the observed effects and he assumes that Newton provided an uniqueness proof of precisely this kind for his mechanics (he gives an outline of the proof; cf. vol. 1, ch. 8, n.13). Knowledge, according to Mill, consists of hypotheses that have been proved in this manner. In *On Liberty*, Mill omits this addition (though there remain traces of it, as in the passage quoted in an abridged form in n.50 below) and gives us a theory of knowledge that is far more realistic than Popper's; cf. the text to n.54 below.

instances or counter-examples, refutations and attempted refutations – in short, criticism – ' while the justificationist lays stress on positive arguments, positive instances, and the like.[50] 'So essential is this [critical] discussion to a real understanding of moral and human subjects that, if opponents of all important truths do not exist, it is indispensable to imagine them and to supply them with the strongest arguments which the most skillful devil's advocate can conjure up', writes Mill on the same point.[51] The theory that survives the debate is of course not free from fault. It dominates the field not because it has been proved, or because it is highly probable, or because it possesses some other desirable feature such as consistency. Even the best scientific theory is usually in a pretty sorry state. What *is* decisive is that its faults are smaller than the faults of the rivals, for the time being. Thus 'the beliefs which we have most warrant for have no safeguard to rest on but a standing invitation to the whole world to prove them unfounded', i.e. to find reasons for rejecting them. 'If with every opportunity for contesting it [a certain hypothesis] has not been refuted' then we shall regard it as better than an alternative which, while supported by all the available evidence 'has not gone through a similar process'. 'If even the Newtonian philosophy were not permitted to be questioned, mankind could not feel as complete assurance of its truth as they do now.'[52] And 'the edifice of our knowledge does not consist of . . . well established truths . . . It consists of largely arbitrary elements . . . socalled hypotheses'.[53] Or, in Popper's repetition '. . . all that can possibly be "*positive*" in our scientific knowledge is positive *only* in so far as certain theories are, at a certain moment of time, preferred to others in the light of our *critical* discussion . . . Thus even what may be called "positive" is so *only* with respect to *negative methods*' (20, starting 'In my view . . .', original italics).

The analogy to Darwin is obvious. Before Darwin it was customary to view organisms as divinely created *and therefore perfect* solutions to the problems of life on this planet. During the nineteenth century one became aware of the numerous mistakes of the life process. This process is not a carefully planned and meticulously performed attempt to realize an aim that has been thought out in advance. It is unreasonable, wasteful, it

[50] *Objective Knowledge*, 20, 32. Mill writes 'It is the fashion of the present times to disparage negative logic – that which points out weaknesses in theory or errors in practice, without establishing positive truths. Such negative criticism – cannot be valued too highly . . . as a means to attaining any positive knowledge or conviction worthy the name.' Cohen, *Philosophy of J. S. Mill*, 236f.

[51] Mill, quoted from Cohen, *Philosophy of J. S. Mill*, 228. Mill does not restrict this method to 'moral and human subjects'. Cf. Cohen, 208 as well as the text below.

[52] Cohen, 209, 207, 208, again 209. This work was written in the heyday of Newtonianism.

[53] Boltzmann, *Populäre Schriften*, ch. 8.

produces an immense variety of forms, *and leaves it to nature* to eliminate the duds. The remaining forms are surprisingly efficient, as if they had been planned with a definite aim in mind. However, there was no aim, there was no 'method of justification', and it is very doubtful whether conscious planning could ever have produced a comparable result.

The analogy does not *solve* methodological problems, it only shows what kinds of solution are *possible*. The world of an organism is a natural world; it acts blindly. The world of a theory is a social world; it is built up by scientists who have to decide what to keep and what to eliminate. Is the decision to be completely arbitrary, is it supposed to proceed according to explicit rules and, if the latter, which rules shall we choose? These are the questions which arise once we start relying on methods of elimination instead of looking for methods of justification.

Mill dispenses with general rules and judges every case on its own merits. He seems to think that the shape of the rival theories and the historical situation in which they compete should have an influence on our judgement, and he seems to believe that our ingenuity will, and should be permitted to react differently to different types of influence: 'The human faculties of perception, judgement, discriminative feeling, mental activity, and even moral preference are exercised . . . in making a choice' and they are 'improved' by such an exercise.[54] Should the improved faculties not be given the chance to make improved judgements on the basis of improved standards?

Popper denies this. He sets up rules of elimination that treat all theories in the same way, independently of the historical situation in which they are judged, independently of any development in our *methodological* imagination that might have taken place. This unhistorical procedure is of course a result of the ideology of the Vienna Circle.

Imre Lakatos has shown that Popper's rules are too restrictive and that they are liable to wipe out science as we know it.[55] He has replaced Popper's rather narrow standards by standards that lie somewhere between Mill and Popper. The standards take account of the shape of the rival theories and of the circumstances in which they compete, *but they do this in a general way*. Methodological imagination is used without being permitted to run loose.[56]

[54] Cohen, 252.

[55] Cf. 'History of Science and its Rational Reconstructions', in *Boston Studies in the Philosophy of Science* VIII, ed. R. C. Buck & R. S. Cohen (Dordrecht, 1971).

[56] Using the terminology of Lakatos we may say that Mill's methodological research programme was progressive. It predicted that any major change of outlook would involve at least two theories, and that the defeated theory would only rarely disappear without a trace; and this prediction is confirmed by every single case study. Mill made no prediction about the shape of the theories before and after the change; for example, he did not say that one was refuted (in the technical sense of Popper) while the other was not, nor did he demand that both theories be falsifiable. Popper's research programme, on the other hand, degenerated from the very beginning though its degeneration was concealed by a tendentious presentation of historical events. According to Popper, theories that participate in a scientific

In addition Lakatos has provided a method for judging not just theories, but also standards, and he has shown in detail how the method can be employed. He has considerably improved the critical inventory of critical rationalism. What has Popper to say about this improvement and about the criticism of his own views from which it proceeds? The answer can be given in a single word: *nothing*.

For although Popper promises a 'full answer to [his] critics' he never even mentions the decisive objections of Kuhn and Lakatos. Nor is it clear who the 'critics' are whom Popper has in mind though, judging from the arguments he offers, they seem to come mainly from the backwoods of Oxford. There is no trace of an answer to more interesting, relevant, and hard-hitting objections. This is a strange procedure for a thinker who more than anyone else has preached the need for a critical discussion and who continues to maintain, in the face of a vast amount of contrary evidence, from general history, as well as from the history of science, that 'what counts in the long run is a good argument' (239). There is only one passage which indicates that Popper has read Lakatos and that he did not like what he saw. I shall now, finally, discuss some of the issues raised in this passage.

14. REFUTABILITY OF PHYSICAL THEORIES

The name index of the book says that Lakatos occurs in a footnote on p. 38. The footnote shows no trace of Lakatos; what it *does* contain is the blunt statement 'that the refutability of Newton's or Einstein's theory is a fact of elementary physics and of elementary methodology', and the equally blunt assertion that Einstein has said so himself: 'Einstein, for example, said that if the red shift effect . . . was not observed in the case of white dwarfs, his theory of general relativity would be refuted' (38, n.5). This is a kind of an answer to Lakatos' detailed criticism of the first statement, only it is not an argument. Besides, both statements can easily be shown to be false. Let us take the alleged Einstein quotation first.

Popper is not the only writer who ascribes to Einstein what Lakatos has called a naive-falsificationist attitude. Herbert Feigl, for example writes: 'If Einstein relied on "beauty", "harmony", "symmetry", "elegance" in constructing . . . his general theory of relativity, it must nevertheless be remem-

competition are falsifiable and a major change of outlook involves a theory, a big falsifying fact and, possibly though not necessarily, a new theory that does not conflict with any existing basic statement. This 'prediction' of Popper's could be adapted to the historical evidence only with the help of degenerating problem shifts. By now this series of shifts seems to have reached a dead end. Lakatos is an improvement not upon *Popper*, but upon *Mill*: he gives an account of the 'fine structure' of major changes where Mill gave only rough outlines and the specifics of *this* account have again been confirmed. The catalyst that leads from Mill to Lakatos is the philosophy of dialectical materialism. The catalyst that is responsible for the Popperian degeneration is the philosophy of the Vienna Circle.

bered that he also said (in a lecture in Prague in 1920 – I was present when I was a very young student): "If the observations of the red shift in the spectra of massive stars don't come out quantitatively in accordance with the principles of general relativity, then my theory will be dust and ashes"',[57] and he has frequently told the story in his lectures. Popper gives no source for his version of the story, and he most likely has it from Feigl. But Feigl's story and Popper's repetition conflict with the numerous occasions when Einstein emphasizes the 'reason of the matter' over and above 'verification by little effects', and this not only in casual remarks, during a lecture (assuming he ever made the remark attributed to him by Feigl), but in writing.[58,59] In 1952 Max Born writes to Einstein as follows: 'It really looks as if your formula is not quite correct. It looks even worse in the case of the redshift [the crucial case referred to by Feigl and Popper]; this is much smaller than the theoretical value towards the center of the sun's disk, and much larger at the edges . . . Could this be a hint of non-linearity?[60,61] Einstein replies: 'Freundlich [the authority reporting the experimental results] does not move me in the slightest. Even if the deflection of light, the perihelian movement or line shift were unknown, the gravitation equations would still be convincing because they avoid the inertial system (the phantom which affects everything but is not itself affected). *It is really strange*

[57] *Minnesota Studies in the Philosophy of Science* v (Minneapolis, 1970), 7.

[58] For instance: 'The external conditions which are set for [the scientist] by the facts of experience do not permit him to let himself be too much restricted, in the construction of his conceptual world, by the adherence to an epistemological system. He therefore, must appear to the systematic epistemologist as a type of unscrupulous opportunist', and must make him believe that he is holding one view when, in fact, he holds another. *Albert Einstein: Philosopher-Scientist*, ed. P. A. Schilpp (Evanston, 1951), 683f.

[59] In 1905–6 Kaufmann tested some consequences of the Lorentz- (and Einstein-) electron and compared them with the rigid electron of Abraham. ('Ueber die Konstitution des Elektrons'. *Ann. Phys*., 19 (1906), 487.) His conclusion was unambiguous and was stated in italics: '*The results of the measurements are not compatible with the fundamental assumption of Lorentz and Einstein*.' Lorentz's reaction: 'It seems very likely that we shall have to relinquish this idea altogether.' (*Theory of Electrons*, 2nd edition (New York, 1952), 213.) Ehrenfest: 'Kaufmann demonstrates that Lorentz' deformable electron is ruled out by the measurements.' ('Zur Stabilitätsfrage bei den Bucherer-Langevin Elektronen', *Phys. Z.* 7 (1906), 302). Poincaré's reluctance to accept the 'new mechanics' of Lorentz can be explained, at least in part by the outcome of Kaufmann's experiment. Cf. *Science and Method* (New York, 1954), book 3, ch. 2, section 5, where Kaufmann's experiment is discussed in detail, the conclusion being that the 'principle of relativity . . . cannot have the fundamental importance one was inclined to ascribe to it'. Cf. also St Goldberg, 'Poincaré's Silence and Einstein's Relativity', *Br. J. Hist. Sci*., 5 (1970), 73ff, and the literature given there. Einstein himself regarded the result as 'improbable because its basic assumption, from which the mass of the moving electron is deduced, is not suggested by theoretical systems which encompass wider classes of phenomena', *Jahrb. Radioaktiv. Elektron.* 4 (1907), 439. Cf. also Einstein's letters to M. Besso and C. Seelig as quoted in G. Holton, 'Influences on Einstein's Early Work in Relativity Theory', *Organon*, 3 (1966), 242, and C. Seelig, *Albert Einstein* (Zürich, 1960, p. 271).

[60] *Born–Einstein Letters* (New York, 1971), 190, dealing with Freundlich's analysis of the bending of light near the sun and the redshift.

[61] Non-linearity was a favourite pastime of Born's.

that human beings are normally deaf to the strongest arguments while they are always inclined to overestimate measuring accuracies' (my italics).[62] How is this conflict between Feigl's story and the written evidence for Einstein's attitude to be explained?

It cannot be explained by assuming a *change* in Einstein's point of view (e.g. Einstein was a Popperian when he invented the general theory of relativity, and became a metaphysician only later on). His disrespectful attitude towards observation and experiment *was there from the very beginning*, as can be seen from his early writings and his attitude in the face of experimental difficulties.[63] The conflict might be explained either as an error on Feigl's part, or, more probably, as an instance of Einstein's 'opportunism'.[64] At any rate, building an argument against Lakatos on what Einstein has allegedly said on the occasion of a popular lecture means building on sand (this is not the only occasion when Popper's reference to authoritative pronouncements by scientific stars turns out to be somewhat less than reliable).

The assertion however 'that the refutability of Newton's and Einstein's theory is a fact of elementary physics and of elementary methodology' should be compared with Lakatos' very clear and very detailed arguments to the contrary.[65] Nothing needs to be added to his account, except perhaps the remark that it is strong enough to withstand even Popper's *ipse dixit*. To explain the situation to the reader who is not familiar with the work of Lakatos, let me use the beloved example

All ravens are black — Sentence *R*

Given a black raven, nothing can be said about either the truth or the falsity of *R*. Given a white raven, we may infer that *R* is false. It is this elementary asymmetry that is the basis of Popper's argument.

The asymmetry remains even if the *identification* of black ravens or of white ravens should turn out to be an unending process, never leading to complete certainty. For it is not asserted that negative instances are more easily *identified* than positive instances; what is asserted is that having identified a negative instance, and having formulated the identification in a 'basic statement', *we can draw an inference* as to the truth-value of *R* (*R* is false), while identification of a positive instance does not permit us to draw such an inference. So far Popper's case against his opponents.[66]

[62] Letter of 12 May 1952, *ibid.*, 192.
[63] Cf. n.59.
[64] Cf. the quotation in n.59. Serious epistemologists always get into trouble when confronted with witty and mobile thinkers such as Einstein (or Galileo). They regard casual statements as descriptions of basic philosophy and assume that basic philosophy is always expressed in neat and clear statements.
[65] Cf. his classic 'Falsification and the Methodology of Research Programmes' in *Criticism and the Growth of Knowledge*, ed. I. Lakatos & A. Musgrave (Cambridge, 1970).
[66] The argument refuted in this paragraph is due to Ayer and Juhos.

To open the case against *him*, we need only consider that not every white raven will refute *R*. A raven that has been painted white, or a raven that has fallen into a bag of flour are both perfectly white – but we shall hardly be inclined to reject *R* on their account. The reason is that *R* does not simply assert that all ravens are black, it asserts that they are 'intrinsically' black, which is quite compatible with their being white as a result of special circumstances. The question is: what are 'special circumstances'?

Now it is clear that we cannot specify 'special circumstances' once and for all and define colour terms accordingly. Or rather, it would be very unwise to do so, for any such speculation would freeze a particular stage of knowledge (a particular set of empirical hypotheses) concerning intrinsic and extrinsic colour and hide the assumptions involved under definitions and usages of words. Assume, for example, that we exclude *external* influences and specify: colours which are due to external influences are not intrinsic colours. Then we may still discover that the blackness of the common crow is the result of very complex external factors and must therefore be discarded. Or we may find that the external factors worked via an internal mechanism and thus gave rise to intrinsic black. We may also discover a 'hierarchy' of 'internal' mechanisms, some more fundamental (genes), some less so (temporary metabolic disturbances), and so further refine the notion of an 'intrinsic' colour. Result: the nature of the special circumstances whose absence or presence permits or prevents refutation depends on the results of empirical research; it changes with these results and so do our ideas concerning the intrinsic properties, whose absence or presence is being asserted. This means, of course, that anticipating the results of future research, we may uphold *R* in the face of a raven whose whiteness has been ascertained to such an extent that 'Alan (being the name of the raven) is white' becomes an accepted basic statement. We confront 'all ravens are black' with 'Alan is white' *without* rejecting the former sentence, and without even being able to say whether, and under what circumstances, it will ever be possible to reject it. Popper hardly ever considers this eventuality, and when he does he dogmatically denies its existence. Lakatos, on the other hand, has reminded us that the *ceteris paribus* clause covers cases of precisely this kind. Considering that we are dealing with intrinsic colours, both in *R* and in the basic statements, we do not simply confront the white Alan with *R*, we confront the white Alan with *R* and the *ceteris paribus* clause, and so we may retain *R*, regard the *ceteris paribus* clause as refuted and begin research concerning the circumstances responsible for the refutation. And since such research may lead to a new view of intrinsic properties and thereby to a redefinition of crucial terms in *R*, we realize that even an innocent-looking statement such as *R* is a *research programme* whose adaptable machinery is not formulated in detail but hidden in the rules of usage of its main descriptive terms.

The situation provides a further argument for the need of alternatives in the process of testing. A white raven refutes *R* only if its whiteness is intrinsic (and if the *ceteris paribus* clause is retained). We therefore need some insight into the causal mechanism that brought the whiteness about, i.e. we need a theory of colour production in animals (in the plumage of birds). The theory must entail intrinsic whiteness in some ravens, i.e. it must conflict with *R*. This is how the need for alternatives can be established even in trite cases such as 'All ravens are black'. This closes the case against the naive assumption that there exist statements whose refutability is 'a fact of elementary methodology'.

15. CONCLUSION

At first reading, Popper's book makes a tremendous impression. This impression has blinded some of its already not too clearsighted reviewers.[67] But look at the reasons given and the doctrines proposed, consider the progress made in all fields, and especially in methodology, since the publication of Popper's *opus magnum*, considers its predecessors, such as Mill and other thinkers of the nineteenth century, and you will be surprised to see how difficult it is to find a moderately acceptable argument, how often blunt assertions, equivocations, rhetorical questions take the place of rational discourse, how little more recent discoveries are taken into account and how small the difference is between the valuable parts of his book and the views of his predecessors. We are not too far from the truth when saying that with *Objective Knowledge* Popper's research programme has entered its degenerating phase.

[67] Why is it that people are so easily impressed by empty profundities? And why is it that Popper, who was once in the forefront of those who opposed them by humour and by argument, has now ceased to be capable of using either?

10

The methodology of scientific research programmes

SUMMARY

The historical studies on which the following essay comments[1] use the idea of a research programme (explained in the first essay by Lakatos) and define two types of relation between research programmes and the evidence. Let me call these type *A* and type *L* respectively. They examine episodes where one research programme, R'', replaces another research programme, R' (or fails to be replaced by it), i.e. R'' is made the basis of research, argument, metaphysical speculation by the great majority of competent scientists. The authors find that the relation of R'' to the evidence is always of type *L* while that of R' is of type *A* (other circumstances being present when this is not the case). If the historical analysis is correct this is an interesting *sociological law*. The authors do not present their results in such terms, however. Making *A* and *L* part of a *normative methodology*, they claim to have shown that the acceptance of R'' was *rational* while the continued defence of R' would have been *irrational*, and they express this belief of theirs by calling research programmes exhibiting relation *L* to the evidence *progressive* research programmes, while research programmes which stand in relation *A* to the evidence are called *degenerating*. They also claim that such judgements are *objective*, independent of the whims and subjective convictions of the thinkers who make them. Using such a normative interpretation of their sociological results they also claim to possess *arguments* for and against research programmes. For example, they would say that today most versions of environmentalism degenerate and that it is irrational to continue working on them. Fortunately this puritanical superstructure of the otherwise excellent sociological studies need not be taken seriously. The reason is that the superstructure is arbitrary, or 'subjective' in at least five different ways. (i) The basic philosophy behind the normative appraisals makes modern science the source of the standards without giving reasons for this choice; (ii) despite all its praise for modern science it uses a streamlined version of it without (*a*) making the principles of

[1] Collected in C. Howson (ed.) *Method and Appraisal in the Physical Sciences* (Cambridge, 1976). Page references and references to individual essays will be to this volume. The present chapter was originally published as the concluding essay in the Howson volume, under the title 'On the Critique of Scientific Reason'.

streamlining explicit and without (*b*) arguing for them; (iii) the standards that are obtained via the arbitrary steps (i) and (ii) are not strong enough to praise any action as 'rational' or condemn it as 'irrational', which means that such judgements are without any support from the arbitrarily accepted methodology; (iv) in some of the essays research programmes are selected in an idiosyncratic manner, the purpose being to make the general philosophy appear true (not that such truth would be of any use; see (iii) above); (v) the attempt to show that competent scientists always act 'rationally' is not applied to all scientists but only to those whose actions seem to fit into the general methodology (for 'seem' see again (iii) above). The superstructure which is subjective in the five ways just enumerated is supposed to guide scientists, while the case studies are to show that the guide has some substance – he is not merely a philosopher indulging in abstract dreams of law and order. I argue that the alleged substantiality is moonshine and that one can reject the standards just as arbitrarily as they have been introduced. To sum up: in the essays in the Howson volume we have (*a*) the discussion of certain sociological regularities; (*b*) the proposal of arbitrary standards which have no practical force; (*c*) the insinuation that the regularities are not merely factual, but are features of rationality and that they lend support to the standards, and give them the authority they need to be accepted as guides of research. (*a*) may be accepted with the caution we extend to any new 'discovery' in sociology; (*c*) must be rejected (and with it the tendentious terminology used in the presentation of the results of (*a*)); (*b*) may be accepted, or rejected, depending on mood, the weather, etc. Environmentalists, however, may continue on their path, for no argument has been raised against their enterprise.

1. *Two fundamental questions, only one of them examined by the methodology of research programmes*

There are two questions that arise in the course of any critique of scientific reason. They are:

(i) *What is science?* How does it proceed, what are its results, how do its procedures, standards and results differ from the procedures, standards and results of other enterprises?

(ii) *What's so great about science?* What makes science preferable to other forms of life, using different standards and getting different kinds of results as a consequence? What makes modern science preferable to the science of the Aristotelians, or to the ideology of the Azande?

Note that in trying to answer question (ii) we are not permitted to judge the alternatives to science by scientific standards. We are now *examining* these standards, we are *comparing* them with other standards rather than making them the *basis* of our judgements. Azande results must be judged by Azande standards and the question is: are these results and these standards preferable to science, or are they not? And if they are not, then, what are the reasons for the deficiency?

Questions (i) and (ii) arise with all abstract concepts. We can ask them about truth, knowledge, beauty, goodness, and so on. In the history of thought answers to question (ii) are often taken for granted. For example, it is taken for granted that truth is something quite excellent and that all we need to know are the detailed features of this excellent thing. This means that one starts with a *word* and uses the enthusiasm created by its *sound* for the support of questionable ideologies (cf. the Nazis on freedom)

There are at least two ways of dealing with question (i). We may use the method of an *anthropologist* who examines the behaviour and the ideology of an interesting and peculiar tribe. In this case statements such as 'science proceeds by induction' are *factual* statements of the same kind as statement describing how a particular tribe builds houses, how the foundation is laid, how the walls are erected, what rites accompany the procedure, and so on. On the other hand we may consider *ideal demands* and examine their consequences. Such a procedure is only loosely connected with actual (scientific) practice, and it may be entirely divorced from it. This applies to many investigations dealing with the 'logic' of science. Occasionally the difference is noticed, but emphasized as an advantage: actual science has not yet achieved the purity of an enterprise that agrees with the demands of a so-called 'rational' enterprise; it must be 'reconstructed' and the reconstruction, obviously, will be different from the real thing. Of course, nobody can say whether a reconstruction, *when inserted into the historical surroundings that gave birth to actual science*, will produce comparable results. It may not give any results at all. (Who would expect that one can climb Mount Everest using the 'rational' steps of classical ballet?) We need the anthropological method to find out whether reconstruction improves science, or whether it turns it into a useless though perfect adornment of logic books. The procedure of the anthropologist therefore takes precedence over the procedure of the logician.

There is a third way in which science and, for that matter, any practice can be examined. Considering the standards and the aims of a certain form of life we may ask whether the practice agrees with the standards and whether it leads up to the aim. In this case we compare the results of an anthropological inquiry (the practice, the aims, the standards which have been found to constitute the form of life) with what we know, or think we know about the laws of nature and man's relation to them. For example, we may point out that a certain way of building houses is not very efficient and that houses built in this way cannot last very long. Or we may point out that induction does not get us very far, and does not provide certainty. The first criticism assumes (*a*) that the builder wants to build a solid house and (*b*) that our knowledge of building houses is at least as good as the knowledge of the culture we are examining. That (*a*) is not always satisfied can be seen from our own civilization which relies on obsolescence. And as regards (*b*), it suffices to remind the reader of recent discoveries which show that 'primitive' procedures are often superior to their scientific rivals.[2] The second criticism, the criticism of induction, assumes that its users wants to get very far (they do not) and that the classes corresponding to universal properties have misleading subclasses[3] (if there are no misleading subclasses then induction will succeed despite the alleged invalidity of the inference from the particular to the general). We see: a true Critique of Scientific Reason cannot take anything for granted. It must examine the most obvious assumptions.

2. *The excellence of science is* assumed, *it is not argued for. The same is true of the standards proposed by the methodology of research programmes. These standards are obtained by an analysis of modern science. Their excellence is therefore again assumed, it is not argued for. There is not a single argument to show that they are better than the standards that underlie the practice of magic*

Such a critical attitude is only rarely found among philosophers of science. Scientists have by now gone very far in the revision of basic cosmological ideas and they have come up with some amazing ideas (subject dependence of the physical world; synchronicity in addition to causality; telekinesis; non-sensory information gathering by plants and ability to recognize individuals; non-causal reaction of deep sea organisms to the position of the sun and of the moon; artificial character of the first satellite of Mars; existence of an international astronomy at 30,000 B.C. and so on). There is no longer any antagonism between the most advanced parts of science and ancient points of view which have degenerated because of scientific warfare. Ancient myths are reconsidered, brought into testable form, examined. Surprising and revolutionary results have been obtained, in the Soviet Union,[4] in China,[5] in the United States.[6] Speculation on the frontiers of knowledge is often indistinguishable from mythmaking and does not follow any easily recognizable method. There may be law and order in some domains; there is absolutely no law and order in others. It is true that the great majority of scientists is still quite hostile towards such mobility (the National Science Foundation, for example, refuses to support

[2] Cf. my *SFS*, 100ff.

[3] Cf. section 6, on the cosmological criticism of methodologies.

[4] S. Ostrander & L. Schroeder, *Psychic Discoveries behind the Iron Curtain* (Englewood Cliffs N.J., 1970), as well as the literature in Thelma Moss' contribution to *Psychic Explanation, A Challenge for Science*, ed. A. D. Mitchell (New York, 1974).

[5] Cf. the immense literature on traditional medicine in China as well as my brief account in ch. 4 of my *AM*.

[6] Cf. the reports and literature in Mitchell, *Psychic Explanation*.

the most interesting research on plant communication that is being carried out by some members of the Stanford Research Institute in Menlo Park). This is a familiar phenomenon, explainable by prejudice and an anxious commitment to the status quo. On the other hand, one would have thought that philosophers of science, being aware of such developments and being less impressed by scientific orthodoxy than their specialist colleagues, might have developed a suitable philosophy, providing additional stimuli for speculation.

This has not been the case. Quite the contrary: most philosophers of science still seem to be living in Victorian times when only a few clouds were dimly perceived on a distant horizon. Their craving for orderliness easily exceeds that of the most systematic scientist and approaches that of a catatonic. They have a strong faith in the basic orderliness of science, they have a strong faith in the excellence of (non-dialectical) logic (and this despite the many open problems one finds even in this discipline), and they spend their lives trying to find a point of view that enables them to uphold both kinds of faith. In this they often succeed, for 'science' is for them a particular logical system, or set of systems, rather than the historical process usually designated by that name, and 'logic' is a very simple and dull part of that discipline, a kind of pidgin logic. 'Problems of science', however, are the internal problems of the chosen system, or set of systems, illustrated with the help of bowdlerized examples from science itself.[7] Kuhn has shown the dream-like character of the whole enterprise. The essays under review and the methodology of research programmes that forms their background are an attempt to move from dream to reality *without* any loss of logic and reconstruction. Let us inquire to what extent the attempt succeeds.

We see at once that question (ii) remains unanswered. It is of course assumed that science is vastly better than any other research programme of comparable scope and generality. But we do not find a single reason in favour of this assumption. Occasionally the assumption enters a detailed argument concerning some different matter, apparently lending it ad-

[7] According to Nelson Goodman, in 'A World of Individuals', quoted from *The Philosophy of Mathematics*, ed. P. Benacerraf & H. Putnam (Englewood Cliffs, 1964), 209, the inventions of the scientists 'become raw material for the philosopher whose task is to make sense of all this . . . in understandable terms'. Considering that logicians are only rarely capable of following scientists on their flights of fancy this would indeed seem to be 'the task' of 'the philosopher' – only Goodman is not that modest. If *he* does not comprehend a thing, then the thing uncomprehended is *intrinsically obscure* and must be 'clarified', i.e. it must be translated into a language which he understands (pidgin logic, in most cases). If *he* understands a language, then the language is intrinsically clear and must be spoken by everyone. This is also how the demand for reconstruction arises. Logicians cannot make sense of science, but they can make sense of logic and so they stipulate that science must be presented in terms of their favourite logical system. This would be excellent comedy material were it not the case that by now almost everyone has started taking the logicians seriously.

ditional force when all we have is a dogmatic and ritual reassertion of the greatness of science.[8]

Thus John Worrall in a position paper on critical rationalism[9] compares the measures which Marxists use to get rid of *prima facie* refuting instances with the measures used by scientists, and he asserts that the former do not lead to any increase of content while the latter do. Had he examined the matter with the care he has spent on his story of Young and Fresnel he would have come to a different conclusion.[10] But let us assume he is right; what follows? We can infer that Marxism does not agree with the standards of science as reconstructed by critical rationalists. We cannot infer that Marxism is *inferior* to science unless we have reasons for the standards

[8] It is surprising to realize how difficult it is to see science in perspective. Carl Sagan, surely one of the most imaginative scientists alive, warns us not to unduly restrict the possibilities of life, and he mentions various types of 'chauvinism' (oxygen chauvinism: if a planet has no oxygen, then it is uninhabitable; temperature chauvinism: low temperatures such as those on Jupiter and high temperatures such as those on Venus make life impossible; carbon chauvinism: all biological systems are constructed of carbon compounds) which he regards as unwarranted (C. Sagan, *The Cosmic Connection* (New York, 1973), ch. 6). He writes (179) 'It is not a question of whether we are emotionally prepared in the long run to confront a message from the stars. It is whether we can develop a sense that beings with quite different evolutionary histories, beings who may look far different from us, even "monstrous" may, nevertheless, be worthy of friendship and reverence, brotherhood and trust.' Still, in discussing the question of whether the message on the plaque of *Pioneer 10* will be comprehensible to extraterrestrial beings he says that 'it is written in the only language we share with the recipients: science'. (Cf. 217: messages to extraterrestial beings 'will be based upon communalities between the transmitting and the receiving civilization. Those communalities are, of course, not any spoken or written language or any common, instinctual encoding in our genetic materials, but rather what we truly share in common – the universe around us, science and mathematics.') In times of stress this belief in science and its temporary results may become a veritable mania making people disregard their lives for what they think to be the truth. Cf. Zhores A. Medvedev's account of the Lysenko case in *The Rise and Fall of T. D. Lysenko* (New York, 1969).

[9] 'Criteria of Scientific Progress, A Critical Rationalist View' (mimeographed, 1975), 2/21ff.

[10] Critical rationalists take great care to show that *prima facie* disreputable procedures in science, when looked at in detail, turn out to be quite acceptable (cf. Zahar on the Lorentz–Fitzgerald contraction, or Worrall on the fate of Young's version of the wave theory). They also know that there are good scientists and bad scientists and that the procedures of the former are not discredited by the errors of the latter: no one would abandon science because it contains complementarity. The attitude towards Marxism, or astrology, or other traditional heresies is very different. Here the most superficial examination and the most shoddy arguments are deemed sufficient. Worrall uses some Marxist interpretations of events in Hungary to discredit the whole approach but without saying what the interpretations are, who has put them forth, and where they can be found. K. R. Popper (*The Open Society and its Enemies* (Princeton, 1945) II, 187ff) mentions the hypothesis of colonial exploitation as a perfect example of an *ad hoc* hypothesis although it is accompanied by a wealth of novel predictions (the arrival and structure of monopolies being one of them). And whoever has read Rosa Luxemburg's reply to Bernstein's criticism of Marx or Trotsky's account of why the Russian Revolution took place in a backward country (cf. also Lenin 'Backward Europe and Advanced Asia' in *Connected Works* (London, 1968), IXX, 99ff) will see that Marxists are pretty close to what Lakatos would like any upstanding rationalist to do, though there is absolutely no need for them to accept his rules. After all, all *he* can say in favour of these rules is that the elite of some enterprise he loves *sometimes* sticks to the rules. (See below.)

which are independent of the fact that they are part of science. No such reasons are found in the philosophy of research programmes. Quite the contrary: it is explicitly stated that science is the measure of method and that good method is the method practised by the 'scientific elite' (23). This, at least, is the theory defended by Lakatos in the introductory essay to the Howson volume, 'History of Science and its Rational Reconstructions'.

According to Lakatos, methodologies are tested, i.e. either defended or attacked, by reference to historical data. The historical data which Lakatos uses are '"basic" appraisals of the scientific elite' (30) or 'basic value judgements' (30) which are *value* judgements about *specific* achievements of science; for example, 'Einstein's theory of relativity of 1919 is superior to Newton's celestial mechanics in the form in which it occurs in Laplace.' For Lakatos such value judgements (which constitute what he calls a 'common scientific wisdom') are a suitable basis for methodological discussions because they are accepted by the great majority of scientists: 'while there has been little agreement concerning a *universal* criterion of the scientific character of theories, there has been considerable agreement over the last two centuries concerning *single* achievements' (23). Basic value judgements can therefore be used for checking theories about science or *rational reconstructions* of science in much the same way in which 'basic' *statements* are used for checking theories about the world. The ways of checking depend on the particular methodology one has chosen to adopt: a falsificationist will reject methodological rules *inconsistent* with basic value judgements, a follower of Lakatos will accept methodological research programmes which 'represent a *progressive shift* in the sequence of research programmes of rational reconstructions' (30). The standard of methodological criticism thus turns out to be the best methodological research programme that is available at the particular time. So far a first approximation of the procedure of Lakatos.

The approximation has omitted two important features of science.

On the one hand basic value judgements are not as uniform as has been assumed. 'Science' is split into numerous disciplines, each of which may adopt a different attitude towards a given theory, and single disciplines are further split into schools, heresies, and so forth. The basic value judgements of an experimentalist will differ from those of a theoretician (cf. Rutherford, or Michelson, or Ehrenhaft on Einstein); a biologist will look at a theory differently from a cosmologist; the faithful Bohrian will regard modifications of the quantum theory with a different eye than will the faithful Einsteinian. Whatever unity remains is dissolved during revolutions, when no principle remains unchallenged, no method unviolated. In addition there are individual differences: Lorentz, Poincaré and Ehrenfest thought that Kaufmann's experiments had refuted the special theory of relativity and were prepared to abandon relativity in the form in which it had been introduced by Einstein, while Einstein himself retained it because of its comprehensiveness.

On the other hand, basic value judgements are only rarely made for good reasons. Everyone agrees now that Copernicus' hypothesis was a big step forward but hardly anyone can give a halfway decent account of it, let alone enumerate the reasons for its excellence. Newton's theory of gravitation was 'highly regarded by the greatest scientists' (24), most of whom were unaware of its difficulties and some of whom believed that it could be derived from Kepler's laws. The quantum theory which suffers from quantitative and qualitative disagreements with the evidence and which is also quite clumsy in places is accepted not *despite* its difficulties, in a *conscious violation* of naive falsificationism, but because 'all evidence points with merciless definiteness in the . . . direction . . . [that] all processes involving . . . unknown interactions conform to the fundamental quantum law'.[11] And so on. *These* are the reasons which produce the basic value judgements whose 'common scientific wisdom' Lakatos occasionally gives such great weight to.[12] Add to this the fact that most scientists accept basic value judgements on trust, they do not examine them, they simply bow to the authority of their specialist colleagues, and one will see that *common scientific wisdom is not very common and it certainly is not very wise.*

Lakatos is aware of the difficulty. He realizes that basic value judgements are not always reasonable (23, n.8) and he admits that 'the scientists' judgement [occasionally] fails' (35). In such cases, he says, it is to be balanced and perhaps even overruled, by the 'philosophers' statute law' (35). The 'rational reconstruction of science' which Lakatos uses as a measure of method is therefore not just the sum total of all basic value judgements; nor is it the best research programme trying to absorb (or to produce) them. It is a 'pluralistic system of authorities' (35) in which basic value judgements are a dominating influence as long as they are uniform *and* reasonable. But when the uniformity disappears, or when 'a tradition degenerates' (36), then general philosophical constraints come to the fore and enforce (restore) reason and uniformity.

Now I have the suspicion that Lakatos vastly underestimates the number of occasions when this is going to be the case. He believes that uniformity of basic value judgements prevailed 'over the last two centuries' (23) when it was actually a very rare event. (Here he is in the same predicament as Kuhn who assumes that a particular normal science may have lasted for decades when in fact it was a very rare event.) But if that is the case, then his rational reconstructions are dominated either by common sense (16, n.58), or by the abstract standards of the 'philosopher's statute law'. Moreover,

[11] Leon Rosenfeld in *Observation and Interpretation*, ed. S. Körner (London, 1957).

[12] 'Is it not . . . *hubris* to try to impose some *a priori* philosophy of science on the most advanced sciences? . . . I think it is' (35).

he accepts a uniformity only if it does not stray too much from his standards: 'When a scientific school degenerates into pseudoscience, it may be worthwhile to force a methodological debate' (36). This means, of course, that the judgements which Lakatos passes so freely are ultimately neither the results of research, nor parts of scientific practice; they are part of an *ideology* which he tries to lay on us in the guise of a 'common scientific wisdom'. We discover here a most interesting difference between the *wording* of Lakatos' proposals and their *cash value*. The methodology of research programmes is introduced with the purpose of aiding rationalism. It is supposed to find historical support for methodological standards. Such standards are to be grounded in history, not in the abstract discussion of abstract possibilities. But the reconstructions which are to provide the historical support are very close to the abstract methodologies supposedly aided and they merge with them at times of crisis. Despite the difference in rhetoric ('Is it not *hubris* to try to impose some *a priori* philosophy of science on the most advanced sciences? . . . I think it is', a sentiment that is forgotten the moment Lakatos enters 'the most advanced' parts of atomic physics, 36, n.131), despite the decision to keep things concrete ('there has been considerable agreement . . . concerning single achievements', 23) Lakatos does not really differ from the traditional epistemologists *except that they argue for their abstract principles while he does not but uses propaganda instead*: he announces that he is going to support his principles by historical research but the results of this research are overruled the moment they conflict with what he thinks a 'rationalist' should do. Here I prefer the procedure of Watkins who in the position paper I have already mentioned (see n.9 above) simply says, 'in a letter to Santa Claus', that the science described by critical rationalists is the science he 'would like to have'.[13] This is not exactly the most sophisticated answer to question (ii) but it is the answer that emerges whenever we examine the procedure of our most recent critical rationalists in somewhat greater detail. It is the answer implicit in the essays that have been collected in this book. Any charge of irrationalism or praise of rationalism which these papers contain is therefore purely subjective, unsupported by either abstract or historical reasons. This will become even clearer as our analysis proceeds.

Let us look at the matter from a slightly different point of view. A 'rational reconstruction' as described by Lakatos comprises concrete judgements about results in a certain domain as well as general standards (we have seen that it is the general standards that really run the reconstruction, and in an arbitrary manner, but let us forget this for the time being). A rational reconstruction as described by Lakatos is rational in the sense that if reflects *what is believed to be a valuable achievement* in the domain. It reflects what one might call the *professional ideology* of the domain. Now even if this

[13] See p. 1/3 of the position paper.

professional ideology consisted of a uniform bulk of basic value judgements only, even if it had no abstract ingredients whatsoever, even then it would not guarantee that the corresponding field has worthwhile results, or that the results are not illusory. Every medicine man proceeds in accordance with complex rules; he compares his tricks and his results with the tricks and the results of other medicine men of the same tribe; he has a rich and coherent professional ideology; and yet no rationalist would be inclined to take him seriously. Astrological medicine employs strict standards and contains fairly uniform basic value judgements, and yet critical rationalists reject its entire professional ideology as 'irrational'. For example, they are not prepared even to consider the 'basic value judgement' that the tropical method of preparing a chart is preferable to the sidereal method or vice versa (the latter opinion being that of Kepler). This possibility of rejecting professional standards *tout court* shows that 'rational reconstructions' *alone* cannot solve the problem of method. To find the right method one must reconstruct the *right discipline*. But what is the right discipline?

Lakatos does not consider the question, and he need not consider it as long as his aim is to find out how post-seventeenth-century science proceeds and as long as he can take it for granted that this enterprise rests on a coherent and uniform professional ideology (we have seen that it does not). But Lakatos and his followers go much further. Having finished their 'reconstruction' of modern science they turn it against other fields *as if it had already been established* that modern science is superior to magic, or to Aristotelian science, and that it has no illusory results. *It is assumed* that question (ii) has already been answered, and that it has been answered in the affirmative. However, there is not a shred of an argument to support this assumption.[14] 'Rational reconstructions' take 'basic scientific wisdom' for granted; they do not show that it is better than the 'basic wisdom' of witches and warlocks. Nobody has shown that only science (of 'the last two centuries', 231) has results that conform to its 'wisdom' while other fields have no corresponding results that conform to their 'wisdom'. What *has* been shown by more recent anthropological studies is that *all* sorts of ideologies and associated institutions produce, and have produced, results that conform to their standards, and other results that do not conform to their standards. For example, Aristotelian science has been able to accommodate numerous facts without changing its basic notions and its basic principles, thus conforming to its own standard of *stability*. We

[14] At this point critical rationalists and followers of the methodology of research programmes usually introduce the criterion of content increase: Aristotle was defeated, and justly so, because he did not conform to this criterion. This assumes (*a*) that Aristotelians *wanted* to conform to the criterion (they did not; see section 5) and (*b*) that the criterion is preferable to, say, the criterion of stability, or to the criterion that the best explanations are *post hoc* explanations. But (*b*) is the assumption under examination.

obviously need further considerations for deciding what field to accept as a measure of method.[15]

Exactly the same problem arises when we consider *particular* methodological rules. It is hardly satisfactory to reject naive falsificationism because it conflicts with the basic value judgements of eminent scientists. Most of these eminent scientists retain refuted theories not because they have some insight into the limits of naive falsificationism, but because they do not realize that the theories are refuted. Besides, even a more rational practice would not be sufficient to reject the rule: universal leniency towards refuted theories may be nothing but a mistake. It certainly is a mistake in a world that contains well-defined species whose properties are only rarely misread by the senses. In such a world the basic laws are manifest, and recalcitrant observations are rightly regarded as indicating an error in our *theories* rather than in our *methodologies*.[16] The situation changes when the disturbances become more insistent and assume the character of an everyday affair. A cosmological discovery of this kind forces us to make a choice: shall we retain naive falsificationism and conclude that knowledge is impossible; or shall we opt for a more abstract and recondite idea of knowledge and a correspondingly more liberal (and less 'empirical') type of methodology? Most scientists, unaware of the nomological–cosmological background of the problem, and even of the problem itself, retain theories that are incompatible with established observations and experiments, and praise them for their excellence. One might say that they make the right choice *by instinct*,[17] but one will hardly regard the resulting behaviour as a measure of method, especially in view of the fact that the 'instinct' has gone wrong on more than one occasion. The cosmological criticism just outlined (omnipresence of disturbances) is to be preferred (and was, as a matter of fact, preferred by Aristotle: see his criticism of the Presocratics).[18]

[15] Watkins, in his 'letter to Santa Claus', points out that the ideal preferred by him is 'really an amalgam of ideas' found in Bacon and Descartes (position paper, 1/4). That may be so, but does not establish its superiority over, say, the ideology of the Aristotelians or of John Dee. Thus, the argument always becomes circular at the most decisive point. Among the contributors to the volume under review only Elie Zahar approaches the problem in a more rational manner. In his account of the Copernican revolution he assumes that all the competitors *shared the same standards* and that the Aristotelians lost because their theories did not conform to these shared standards. This still does not give us an answer to question (ii); all it does is to extend the domain from which the basic value judgements are taken. But it gives a rational account of the victory of the Copernicans *provided* the shared standards had some *force*. This problem will be discussed in the next section.

[16] In such a world the demand for depth (Watkins, position paper, 1/4ff) is unrealistic and cannot be satisfied.

[17] 'Up to the present day it has been the scientific standards, as applied "instinctively" by the scientific elite in *particular cases*, which have constituted the main – although not the exclusive – yardstick of the philosopher's universal laws' (35).

[18] This possibility of *choosing* a methodology on the basis of cosmological considerations shows that there can be different types of science: given fairly clear species with not too many disturbances we may decide to remain naive falsificationists and absorb the exceptions by

To sum up: the methodology of research programmes does not argue for the superiority of science ('of the last two centuries'); it takes this superiority for granted and pretends to use it as a basis for the standards it employs. It does not use it as such a basis because it implies 'a pluralistic system of authority' in which 'the philosophers' statute law' plays an important role side by side with 'common scientific wisdom' ('the philosophers' statute law' is made up of the abstract principles that the methodology of research programmes was supposed to support by appeal to historical facts). Now what is the content of this 'philosophers' statute law', what are the reasons for it, and when does it come to the fore and overrule 'common scientific wisdom'? It comes to the fore 'when the scientists' judgement fails' (35) and *that* occurs whenever there is massive support for degenerating research programmes.[19] Thus the standards, instead of being supported by history, are the *criterion* by which we decide when to accept historical trends and when to reject them. Moreover, the methodology of research programmes does not offer any abstract (philosophical) arguments in their favour (or against alternative standards). The standards are therefore arbitrary, subjective, and 'irrational'. They do not provide *objective* reasons for eliminating Marxism, or Aristotelianism, or Hermeticism, or for attacking new developments in the sciences. They merely indicate what critical rationalists would 'like to have' at this stage of the development of their ideology.

But the situation is even worse. So far I have argued that Imre Lakatos has *not* provided *any objective reasons for accepting the standards*; he has not shown them to be rational in any sense of the word he is prepared to accept. I shall now argue that *the standards have no force either*; they are too weak to condemn any action as 'irrational'. It follows that an author who uses puritanical language of this kind – and the authors of the essays under review use it rather frequently – either subscribes to a rationality theory different from Lakatos', or else he is content with rhetorical flourishes, unconnected with any argument. In the latter case he gives us an interesting sociological study and uses it as a club for forcing people to accept standards which are very different from those he pretends to defend. Let us

methods such as monster-barring, or various means of adaptation; but we may also decide to use basic laws for the explanation of *all* events and so become research programmists. Aristotle made the first decision, Galileo as seen by some thinkers made the second. Thus the idea that there can be only one science – one physics, one biology, one chemistry – which is found even among so-called dialectical materialists (cf. Medvedev, *Rise and Fall of Lysenko*, 133, 247) is again but a result of insufficient analysis.

[19] According to Lakatos it seems that modern particle physics represents a degeneration. He also speaks of the development of 'new bad traditions' such as modern sociology, psychology, social psychology. These traditions are indeed *new*. But they are *bad* only if the standards of science 'of the last two centuries' have been shown to be good. They are *assumed* to be good; this much is sure. But there is no argument to support this assumption and, besides, the standards are overruled whenever they seem to conflict with the house philosophy of the methodology of research programmes.

now examine the assertion that the standards which Lakatos recommends have no force to condemn any action as irrational.

3. *Nor are the standards strong enough to praise individual actions as 'rational' or condemn them as 'irrational'. All that can be said is that the actions have taken place, and that they have certain features*

When a theory enters the scene, it is usually somewhat inarticulate; it contains contradictions; the relation to the facts is unclear; ambiguities abound; the theory is full of faults. However, it can be developed, and it may improve. The natural unit of methodological appraisals is therefore not a single theory, but a succession of theories, or a *research programme*; and we do not judge the state in which a research programme finds itself at a particular moment; we judge its history, preferably in comparison with the history of rival programmes.

According to Lakatos, the judgements are of the following kind: 'A research programme is said to be *progressing* as long as its theoretical growth anticipates its empirical growth, that is as long as it keeps predicting novel facts with some success . . .; it is *stagnating* if its theoretical growth lags behind its empirical growth, that is, as long as it gives only *post hoc* explanations of either chance discoveries or of facts anticipated by, and discovered in a rival programme' (11). A stagnating programme may *degenerate* further until it contains nothing but 'solemn reassertions' of the original position coupled with a repetition, in its own terms, of (the successes of) rival programmes (16). Judgements of this kind are central to the methodology Lakatos wishes to defend. They *describe* the situation in which a scientist may find himself. *They do not yet advise him how to proceed.*

Considering a research programme in an advanced state of degeneration one will feel the urge to abandon it, and to replace it by a more progressive rival. This is an entirely legitimate move. *But it is also* legitimate to do the opposite and to retain the programme. For any attempt to demand its removal on the basis of a rule can be criticized by arguments almost identical with the arguments that eliminate say, naive falsificationism: if it is unwise to reject faulty theories the moment they are born because they might grow and improve, then it is also unwise to reject research programmes on a downward trend because they might recover and attain unforeseen splendour (the butterfly emerges when the caterpillar has reached its lowest state of degeneration).[20] Hence, one cannot *rationally* criticize a scientist who sticks to a degenerating programme and there is no *rational* way of showing that his actions are unreasonable.

[20] This remark shows that the methodology of research programmes, too, makes certain *cosmological* assumptions concerning the relationship between research programmes and the world.

Lakatos agrees with this. He emphasizes that one 'may rationally stick to a degenerating programme until it is overtaken by a rival *and even after*' (15); 'programmes may get out of degenerating troughs'.[21] It is true that his rhetoric frequently carries him much further, showing that he has not yet become accustomed to his own liberal proposals.[22] But when the issue arises in explicit form, then the answer is clear: the methodology of research programmes provides standards that aid the scientist in *evaluating* the historical situation in which he makes his decisions; it does not contain *rules* that tell him what to do.

However, even this very modest formulation still goes much too far. By speaking of *risks*[23] it assumes that the progress initiated by progressive phases will be greater than the progress that follows a degenerating phase; after all, it is quite possible that progress is always followed by long-lasting degeneration, while a short degeneration (say, 50 or 100 years) precedes overwhelming and long-lasting progress.[24] By speaking of evaluation and using evaluative terms such as 'progressive' and 'degenerating' it assumes that 'progress' is preferable to 'degeneration' both intrinsically and as regards consequences. The second case has just been dealt with. The first case (intrinsic advantage of 'progress') leads back to question (ii). Question

[21] I. Lakatos, 'Falsification and the Methodology of Scientific Research Programmes' in *Criticism and the Growth of Knowledge*, ed. I. Lakatos & A. Musgrave (Cambridge, 1970), 164.

[22] 'I give . . . rules for the "elimination" of whole research programmes' (11); note the ambiguity introduced by the quotation marks. Occasionally, the restrictions are formulated in a different way, by denying the 'rationality' of certain procedures. 'It is perfectly rational to play a risky game', says Lakatos (16); 'what is irrational is to deceive oneself about the risk': one can do whatever one wants to do if occasionally one remembers (or merely recites?) the standards *which, incidentally, say nothing about risks or the size of risks*. Speaking about risks either involves a *cosmological* assumption (nature rarely permits research programmes to behave like caterpillars), or a *sociological* assumption (*institutions* only rarely permit degenerating research programmes to survive), and thus leads to exactly the same conundrum which Hume explained so nicely more than 200 years ago. Lakatos in passing (12) admits the need for such additional assumptions: only they 'can turn science from a mere game into an epistemologically rational exercise'. But he does not *discuss* them in detail and those he takes for granted are very doubtful, to say the least. Take the cosmological assumption I have just mentioned. It is interesting, and it certainly deserves to be studied in greater detail. Such a study, I venture to suggest, would reveal that the research programme corresponding to it is now in a degenerating phase (to see this, one needs only consider anomalies such as the Copernican revolution, the revival of the atomic theory, the revival of assumptions concerning celestial influences; and so on). The sociological assumption, on the other hand, is certainly true, which means that given a world in which the cosmological assumption is false we shall forever be prevented from finding the truth.

[23] Cf. the quotation in the previous footnote.

[24] This is, of course, again a version of Hume's problem. Hume turns up in all methodologies because all methodologies make cosmological assumptions. The naive falsificationist assumes that there are no oceans of anomalies. The conventionalist assumes that the world is built in a simple way. The research programmist assumes that progress, once realized, does not put an end to further progress and that it leads to the truth (after all, a progressive theory may lead us further and further away from the truth; cf. the life of Paphnutius as presented by Anatole France). And so on.

(ii) is unanswered, hence the question of the intrinsic advantage of progress is unanswered too.[25]

Alan Musgrave has written an interesting paper in which he agrees with some of the criticisms voiced above.[26] Lakatos, he writes (M, 15) 'develops an elaborate account of what is good science and what bad, but he refuses (apart from "Thou shalt not lie") to give advice to the scientists'. '"Anything goes" is the position which Lakatos finally adopts' (M, n.53). But a methodology must 'provide advice or directives' (M, 22). This advice and these directives are to be addressed 'to science, . . . to the community of scientists, as a whole' (M, 22). They would 'forbid wholesale persistence with degenerating programmes, or premature mass conversion to a budding one' but permit the individual scientist to go his way: 'we cannot condemn Priestley for his die-hard adherence to phlogistonism; but we could condemn the community of late nineteenth-century chemists had they all done the same'. Musgrave thinks he 'can provide a purely deductive argument' for such directives (M, 23). The argument proceeds from

[25] John Worrall writes (p. 2/29, n.3 of the position paper 'Criteria of Scientific Progress, A Critical Rationalist View' (mimeographed, 1975)): 'A scientist *would* be pronounced "irrational" (or rather mistaken) by the methodology of research programmes if he stuck to the old programme denying that the new programme had any merits not shared by the old one and thus denying that his own programme needed improvement in order to catch up with the new one. It is in such circumstances that we shall begin to suspect the operation of extra-rational motives.' The arguments in the last section, in the text above, and in n.21 show that it is rather this judgement of Worrall's which makes us 'suspect the operation of extra-rational motives' where by 'extra-rational' we mean motives either not in accordance with the standards, or not dictated by them. Assume I have a research programme which degenerates and I am told so by a research programmist. My reply might well be that I am interested in certainty and not in novelty and that I prefer a programme that can incorporate newly found facts without revision, to a programme that constantly upsets basic convictions. When being told that this means I am not being 'scientific' I can reply that the excellence of science is still a matter of debate, that it is *assumed* by my opponent (though discarded by him when it goes counter to his own pet ideology; see above), that it is not supported by argument. (Nor is there any argument to show that non-scientific ideologies are worse than science in addition to being different from it; of course, there is a general *belief* that this is so, and this belief may even be quite reasonable, but what I am now talking about is the ability of the methodology of research programmes to give a *reason* for the belief.) Adopting the point of view of science, I can add that degeneration when taken seriously may be followed by bigger progress than progress and that progress may lead away from the truth. Or is a scientist supposed to be satisfied with temporary spectacles only? Is it enough for him to impress everyone by first predicting, and then discovering, a new planet (e.g. Neptune) *without any implications for the quality of future research*? And, finally, one might comment on the futility of a point of view where a thief can steal as much as he wants, is praised as an honest man by the police and by the common folk alike provided he tells everyone that he is a thief. If *that* is the sense in which the methodology of research programme differs from anarchism (Worrall, position paper, 2/30, n.1), then I am ready to become a research programmist. For who does not prefer being praised to being criticized if all he has to do is to describe his actions in the lingo of a particular school?

[26] A. Musgrave, 'Method or Madness? – Can the Methodology of Research Programmes be rescued from epistemological Anarchism?', in *Essays in Memory of Imre Lakatos*, ed. R. S. Cohen, P. K. Feyerabend and M. W. Wartofsky (Dordrecht, 1976). (Page references are preceded by M.)

the premise that 'science ought to devote energy to investigating unsolved scientific problems'. Now, a 'progressive research programme throws up more unsolved problems than a degenerating one' hence, 'science ought to devote more energy to' progressive programmes than to degenerating ones (M, 24). In reply one can point out, first, and still in accordance with the methodology of research programmes, that every success of a progressive programme is a problem for its degenerating rival, so in the end it will be a degenerating programme that 'throws up more unsolved problems'. Secondly, it is not only the *number* of problems that counts, but also their *quality*. Now it is certainly more difficult to find the right questions than to answer 'problems' that are already spelled out in detail. Again the directive advises us to pay attention to degenerating programmes. Thirdly, letting the individual scientist do the dirty work of improving a degenerating programme prejudges the issue in a very unfair way. Today an individual can only rarely attack, let alone solve, the problems that arise in the course of research. Without computers, without expensive equipment, without the help of colleagues and assistants he is doomed from the start. (Just consider the expenses involved in experiments such as those of Reines, or Weber; and where would general relativity be today had Einstein had to carry the expenses of all the experiments carried out to test it?) Musgrave's directive, and this is my fourth point, is also uncomfortably close to the directive of some politicians who advise us to spend educational funds only on those who are already well educated and to let the less educated fight for themselves, the difference being that the advantages of an educated person are much more obvious than the advantages of a 'progressive' research programme. And my fifth point is (question): why should one prefer programmes which have successfully anticipated experimental discoveries to programmes which have no such record? Does such a preference not indicate an inductivistic prejudice? Musgrave thinks it should not and this is why he speaks of problems rather than of the successful anticipation of novel facts. But why should a research programme that creates lots of problems be preferable to a research programme that creates none? For an Aristotelian, absence of problems is a sign that certainty and agreement with facts has been achieved. Popperians do not like certainty and they reject the moves that help us achieve it. They do not like certainty and they think they have also found arguments to support their dislike: certainty is not part of science, therefore it should be rejected. This is firstly not true (cf. the arguments of early Newtonians and Cartesians) and secondly not sufficient: why should we accept science as a measure of excellence? We have to conclude that Musgrave's rescue manoeuvre does not succeed. Its principles are arbitrary, and they lead to results very different from those envisaged by him.

To sum up this part of the argument: the standards which Lakatos has chosen neither issue abstract orders (such as 'abandon degenerating research programmes') nor do they support general judgements concerning the rationality, or irrationality, of a certain course of action (such as 'it is irrational to support degenerating research programmes'). Such orders and such judgements give way to concrete decisions in complex historical situations. Hence, if the enterprise that contains the standards is to be different from the 'chaos' of anarchism, *then such decisions must be made to occur with a certain regularity*. Taken by themselves the standards cannot achieve the regularity, as we have seen. But psychological or sociological *pressures* may do the trick.

Thus, assume that the institutions which publicize the work and the results of the individual scientist, which provide him with an intellectual home where he can feel safe and wanted and which because of their eminence and their (intellectual, financial, political) pull can make him seem important, adopt a *conservative attitude* towards the standards; they refuse to support degenerating research programmes, they withdraw money from them, they ridicule their defenders, they do not publish their results, they make them feel bad in every possible way. The outcome can easily be foreseen: scientists, who are as much in need of emotional and financial security as anyone else, especially today when science has ceased to be a philosophical adventure and has become a business, will revise their 'decisions', and they will tend to reject research programmes on a downward trend.

This conservative attitude adopted by the institutions is not irrational, for it does not conflict with the standards. It is the result of collective policies of the kind encouraged by the standards. The attitude of the individual scientist who adapts so readily to the pressures is not irrational either, for he again decides in a way that is condoned by the standards. We have thus achieved law and order without reducing the liberalism of our methodology. And even the complex nature of the standards now receives a function. For while the standards do not prescribe, or forbid, any particular action, while they are perfectly compatible with the 'anything goes' of the anarchist who is therefore right in regarding them as mere embroideries, they yet give content to the actions of individuals and institutions who have decided to adopt a conservative attitude towards them. *Taken by themselves* the standards are incapable of forbidding the most outrageous behaviour.[27] *Taken in conjunction* with the kind of conservatism just described they have a subtle but firm influence on the scientist. And this is precisely how Lakatos wants them to be used. Considering a degenerating programme he suggests that 'editors of scientific journals should refuse to publish . . . papers [by scientists pursuing the programme] . . . Research foundations, too, should

[27] For a minor exception which by now seems to have become the only point of resistance of the methodology of research programmes cf. p. 323 of the Howson volume.

refuse money' (16). The suggestion is not in conflict with the standards, as we have seen. Nor can it be used to raise the charge of irrationality against alternative suggestions: measured by the standards of the methodology of research programmes the conservative attitude expressed by the suggestion is neither rational nor irrational. It is an interesting sociological fact – nothing more. *But it is eminently rational* according to other standards, for example, according to the standards of commonsense.[28] This wealth of meanings of the word 'rational' is used by Lakatos to maximum effect. In his arguments against naive falsificationism he emphasizes the new 'rationalism' of his standards which permits science to survive. In his arguments against Kuhn and against anarchism he emphasizes the entirely different 'rationality' of commonsense but without informing his audience of the switch and so he can have his cake (have more liberal standards) and eat it too (have them used conservatively) and he can even expect to be regarded as a rationalist in both cases. Indeed, there is a great similarity between Lakatos and the early Church fathers who introduced conservative doctrines in the guise of familiar prayers (which formed the commonsense of the time) and who thereby gradually transformed commonsense itself.[29]

4. *Using the methodology of research programmes as our theory of rationality we must therefore regard the case studies under review as sociological studies and we must disregard the frequent judgements of rationality or irrationality that occur in them (on the other hand, these judgements may be retained if we adopt a different theory of rationality, for example, Hegel's). But though the aim of rationalizing history is never reached, the attempt to reach it has produced a history that is richer in content and more conceptual than its predecessors*

If we take all these things into consideration it is clear that Lakatos has not succeeded in showing 'rational change' where 'Kuhn and Feyerabend see irrational change' (31–2). A revolution occurs when a new research pro-

[28] 'In such decisions', says Lakatos, referring to decisions such as those leading to a conservative use of standards, 'one has to use one's *common sense*' (16, n.58), as long as we recognize that in doing so we *leave* the domain of rationality as defined by the standards and move to an 'external' medium, or to other standards. Lakatos does not always make the change clear. Quite the contrary: in his attack upon opponents he makes full use of our inclination to regard commonsense as inherently rational and to use the word 'rational' in accordance with *its* standards. He accuses his opponents of 'irrationality'. We instinctively agree with him, quite forgetting that the methodology under debate does not support the judgement and does not provide any reasons for making it.

[29] Using the *psychological* hold which the baptismal confession had over the members of the early Christian Church and taking the non-Gnostic interpretation 'as its self-evident content' (A. Harnack, *History of Dogma* (New York, 1961) II, 26). Irenaeus succeeded in defeating the Gnostic heresy. Using the psychological hold which commonsense has over philosophers of science and taking the conservative interpretation of his standards as *its* self-evident content, Lakatos has almost succeeded in convincing us of the reality of his own law-and-order philosophy and the non-ornamental character of his standards: now as before the best propagandists are found in the Church, and in conservative politics.

gramme has accumulated a sufficient number of successes and the orthodox programme suffered a sufficient number of failures for both to be regarded as serious rivals, and when the protagonists of the new programme proclaim the demise of the orthodox scheme. Seen from the point of view of the methodology of research programmes they do this not just because of their standards – the standards are not strong enough for making such a judgement – but because they have adopted a conservative attitude towards the standards (all this assumes, of course, that both sides use the methodology of research programmes in their deliberations, a matter that is open to considerable doubt[30]). Their orthodox opponents have what one might call a 'liberal' attitude; they are prepared to tolerate a lot more degeneration than the conservatives. The standards permit both attitudes. They have nothing to say about the 'rationality' or 'irrationality' of these attitudes and of the developments initiated by them. It follows that the fight between the conservatives and the liberals and the final victory of the conservatives is not a 'rational change' (32) but a 'power struggle' pure and simple, full of 'sordid personal controversy' (34). It is a topic neither for methodology, nor for the theory of rationality, but for 'mob psychology',[31] or, to use a more traditional term, *it is a topic for the sociology of knowledge.*

Exactly the same is true of the essays that have been collected in the Howson volume. Each of these essays describes a battle between alternative research programmes and the victory of one of them. Each essay reconstructs the battle in terms of the methodology of research programmes, using the standards of this methodology as a basis for the proper application of its evaluative terms. The standards are not strong enough to guide such application. Basing our judgements on them we can only say that one programme was *accepted* while the other *receded into the background*; we cannot add that the acceptance was *rational*, or that a rational development took place. Thus, the essays are interesting contributions to the *history*, or the *sociology* of science no matter how hard the authors try to present them as something different. The reader must not be misled by the frequent and rather assured use of terms such as 'rational' or 'irrational' which suggests that the authors have some deeper insight into the historical process. Using the methodology of research programmes they cannot have such insight, as I have tried to show. Of course, they have their *preferences*; they know what kinds of things they 'would like to have' and they defend these things with religious fervour. But neither they nor the authorities on whom they rely have succeeded in turning this fervour into a rational procedure.

Let us therefore from now on regard the case studies as historical studies and let us evaluate them on that basis. We see at once that they are superior

[30] See section 5.

[31] Lakatos in *Criticism and the Growth of Knowledge*, 178, original italics.

to earlier studies of the same kind. The procedure is always the same. First, a certain historical episode is identified: the Copernican revolution (not in the Howson volume); the Einsteinian revolution; the chemical revolution of Lavoisier and its altercation with the phlogiston theory; the rejection of Young's version of the wave theory of light; the battle between phenomenological thermodynamics and the kinetic theory of heat. Then follow explanations of the episodes that have played a role in the literature. Some of these explanations are mere narratives, with conceptual connections, others are psychological, still others are methodological; they try to show how the events arose, in a 'rational' manner, as a result of the determined use of methodological rules. Among the methodologies surveyed are inductivism, naive falsificationism, conventionalism, and the views of Kuhn. Next comes the demolition of the traditional explanations. This demolition is almost always historical: the explanations omit important facts; they conflict with others (inductivism alone is removed by logical arguments). Finally, we have an account in terms of the methodology of research programmes. Guided by a complex methodology this account is richer and more sophisticated than the alternative accounts. It is history of ideas in the best sense of the word. It is *history* because it deals with facts. It is history of *ideas* because it shows conceptual connections between these facts. It is *sophisticated* history of ideas because it uses a rich inventory of conceptual tools (hard core, protective belt, heuristic, progressiveness, degeneration, monster adjustment, recovery of hidden lemmas, and so on) rather than relying on intuition in all cases. The researcher is eqipped with instruments that aid him on his way and are open to inspection so that he can criticize them and replace them by better instruments. It is true, the wish of the writers to arrive at some 'objective' judgements has made these instruments overly intellectual: a researcher who wants to inquire into motivation, or sociological causes hardly receives any help. Even worse, he is discouraged from putting too much weight on causes of this kind. This explains the rather primitive sociology and psychology of some of the essays and the complete absence of any inquiry into when, how and why certain standards were accepted. But there are definite advantages when one compares the history with what other methodologists have to offer. An inductivist, for example, will consider a theory (or, rather, a 'logical reconstruction' of it), the 'evidence' (which again is a reconstruction of the complex experimental results that govern science), and the relation between the two. He has two abstract elements and he examines them irrespective of the historical surroundings in which they arose. Writing history he is interested in the 'rational' parts of science only which are again the elements and their relation. This is why inductivist history is so arid or, if richer, so lacking in conceptual penetration. A naive falsificationist is not much better, for all *he* wants is *some* evidence (no matter in what historical surroundings) that contradicts the theory. The methodology of research programmes, on the other hand, does not examine theories; it examines sequences of theories connected by hard cores, heuristics and intuitive attitudes not all of which need to be formulated explicitly. Already at this point it goes much deeper into the structure of a theory than do the rivals. Secondly, the methodology of research programmes does not examine research programmes by themselves: it examines them in comparison with other research programmes. So the investigation spreads and must ultimately reach every research programme at the period in question. The whole intellectual scene must be taken into account. In the sixteenth and seventeenth centuries this included theology, Aristotelian physics and metaphysics, magic (Neoplatonism), the philosophy of Paracelsus (which centred around medicine and chemistry), alchemy: all subjects which were studied with care by the great Newton himself. Then we must examine whether the predictions made were novel predictions, or whether they were repetitions of things already known. This means we must examine the way in which the research programme was originally introduced,[32] the expectations of the age, the 'accepted facts' and the relation of these facts to current theory. We must know a great amount of material that belongs to the history of ideas and that is often missing even from rather detailed *historical* accounts. Nor is this material just *aufgerafft*. It is collected with an aim in mind, described in terms adapted to this aim and thus essentially ideational. It is true that the basic scheme lends a certain uniformity to the procedure which in the hands of less gifted writers can introduce an element of boredom. It is also ironical to realize that the aim – to give a 'rational' account of development – is never reached and that we are left with a historical narrative only.[33] But on the way to the aim history has been transformed to such an extent that a slight change in our stan-

[32] In his essay 'Why did Einstein's programme supersede Lorentz's?' Zahar writes 'My redefinition of novelty amounts to the claim that *in order to assess the relation between theories and empirical data, one has to take into account the way in which a theory is built and the problems it was designed to solve.* This new criterion of novelty of facts also implies that the traditional methods of historical research are even more vital for evaluating experimental support than Lakatos had already suggested. The historian has to read the private correspondence of the scientist whose ideas he is studying; his purpose will not be to delve into the psyche of the scientist, but to disentangle the heuristic reasoning which the latter used in order to arrive at the new theory' (219).

[33] Environmentalists therefore need not be intimated by Peter Urbach's 'Progress and Degeneration in the "I.Q. Debate"', *Brit. J. Phil. Sci.*, 25 (1974), 99–135, 235–59. What is shown in these papers is that the relation of environmentalism to the evidence is different from the relation of some versions of geneticism to the evidence. *This is all that is shown* though both terminology ('degenerating', 'progressive') and philosophical insinuations *create the impression* that the one type of relation is better, more 'scientific' than the other. My remarks in the text above make it clear that this is not so. Value judgements of this kind are completely arbitrary and subjective and nobody needs to be intimidated by them. But alas! The propagandistic genius of Lakatos has concocted a mixture of propaganda and sham argument that is only too difficult to unravel and so he will have his way, because of the power of his rhetoric. One might call this the *List der Unvernunft*.

dards, say from research programme standards to Hegelian standards, enables us to read it as a history of reason itself.[34]

5. *Remaining lacunae can be explained by the rationalizing tendency of the authors and their blind acceptance of modern science*

Still, the rationalizing tendencies of the authors who have contributed to the Howson volume, and their assumption of the excellence of science make themselves felt and are responsible for some lacunae even in the historical account. Let me mention two of them.

All the authors assume that the defenders of rival research programmes (and perhaps even the onlookers) are influenced by, and are proceeding in accordance with, the methodology of research programmes. They use this methodology as a measure of good science and they behave in accordance with the advice given in the standards. It has emerged that the standards are incapable of giving any advice, but let us now disregard this drawback. Let us assume that the standards do indeed favour progressive programmes (they don't just *state* their progressiveness) and condemn degenerating programmes. It is then assumed by the authors that all good scientists distribute praise and blame in accordance with the methodology of research programmes. They may not be able to give an account of the *principles* of this methodology, but they still proceed as if they held such principles to be the basis of research.

Now this assumption may have some plausibility in periods of peace and uniformity which are most likely ruled by a single methodological framework. It becomes improbable when we are dealing with developments such as the Copernican revolution, or the rise of twentieth-century science, especially quantum theory. Lakatos has emphasized that his standards are based on outstanding science of 'the last two centuries' (23) and he has thereby conceded the possibility of different standards *before* this period. But in his joint paper with Zahar on the Copernican revolution he contends that 'Copernicus' and Kepler's and Galileo's adoption of the heliocentric theory' are 'rationally explainable'.[35] 'The Copernican Revolution', they write,[36] 'became a great scientific revolution . . . simply because it was scientifically superior' where 'scientifically superior' means, of course, superior in terms of the methodology of research programmes. Even Aristarchus's claim should have been taken seriously because 'the geocentric programme had already heuristically degenerated'.[37] Similarly Peter Clark in summarizing the result of his essay says 'It was the degeneration of these attempts to provide a foundation for thermodynamics in terms

[34] Cf. the introduction to Hegel's *Geschichte der Philosophie* (Frankfurt, 1971).
[35] I. Lakatos & E. Zahar, 'Why did Copernicus's Research Programme supersede Ptolemy's?' in *The Copernican Achievement*, ed. R. Westman (Berkeley and Los Angeles, 1976), 380.
[36] *Ibid.*, 31. [37] *Ibid.*, 372, n.53.

of some deeper theory, compared with the empirical progress of the phenomenological programme, which led to the elevation of thermodynamics into a "paradigm" of great science' (44). This again assumes that the moves of contemporary scientists were motivated by an implicit understanding of the methodology of scientific research programmes. Is this assumption correct?

I do not think it is.

In the case of the Copernican revolution we know that the methodology of research programmes, if it was used at all, was not the only methodology in existence. One influential methodology was connected with the slogan of *saving the phenomena*. The slogan presupposes a distinction between *basic physics* and *auxiliary assumptions*. Basic physics describes the processes that are expected to occur in this world, auxiliary assumptions link the processes to the phenomena. In Aristotle basic physics is decidedly empirical, much more so than modern science could ever aspire to be. It starts from 'phenomena' which are either observed facts or assumptions of commonsense,[38] develops terminology for their description and principles for their explanation[39] and incorporates 'new facts' in a 'degenerating' manner. This is regarded as an advantage: the fact that theory and phenomena are related in a manner described as 'degenerating' by Lakatos shows the truth of the principles and the adequacy of the terminology used. The auxiliary assumptions are eventually separated from basic theory and assembled into various disciplines (astronomy, optics, mechanics, and so on). The task of these disciplines is to save the phenomena, not to give a physical account of the processes (motions) that created them.[40] Attempts to *re-absorb* them into basic theory in a 'degenerating' way continue throughout the Middle Ages.[41] They continue with Copernicus who tries to find an arrangement of circular movements that agrees with the basic

[38] G. E. L. Owen, 'Tithenai ta Phainomena' in *Aristote et les problèmes de la méthode* (Louvain, 1961).

[39] For the role of principles in Aristotle cf. H. Wieland *Die Aristotelische Physik* (Heidelberg, 1970). Wieland makes it clear how principles in Aristotle are designed with the explicit purpose of achieving 'degenerating' adaptations of facts (or course, he does not use this terminology).

[40] Originally, in Aristotle and his immediate successors the task is to *give an account of phenomena* in terms of basic physics which in turn must be constructed in such a way that the phenomena can be accounted for. Later on basic physics is taken for granted and phenomena must be explained in its terms. This is how the idea of *saving* the phenomena (rather than giving an account of them) arises. Cf. F. R. Krafft 'Der Mathematikos und der Physikos – Bemerkungen zu der angeblichen Platonischen Aufgabe, die Phaenomene zu retten', in *Alte Probleme – Neue Ansätze. Drei Vorträge von Fritz Krafft* (Mainz, 1965). An account of phenomena deals with the nature of things. The auxiliary assumptions that are used to save the phenomena have no such pretensions. The distinction is prepared by *Physics* B 2.

[41] The attempts start with Ptolemy's *Planetary Hypotheses*, are continued by Arab astronomers of the eleventh and twelfth centuries who demand a realistic account of planetary motions and last until the sixteenth century when the system of Alpetragius is taken up by Purbach. Cf. P. Duhem, *Le Systeme du Monde* (Paris, 1915) II, 130ff.

Aristotelian philosophy[42] to such an extent that it can again be regarded as an account of real motions leading to a coherent *system* of the world rather than as a set of independent *hypotheses* for the calculations of planetary phenomena. Neither the tradition of saving the phenomena nor the attempt to absorb the mathematical devices of this tradition into basic theory look for novel facts or novel explanations of existing facts in the sense of the methodology of research programmes. The former cannot be found for all facts are related to the same basic principles and formulated in their terms, the latter cannot be admitted for that would deny the existence of a stable basic theory. To a certain extent this is true even of Copernicus himself (though not of Rheticus, and certainly not of Kepler[43]). But if that is the case then the revolution that was started by Copernicus' hypothesis was much more dramatic than a transition, *within the methodology of research programmes*, from one research programme to another. It brought in new standards and thus constituted a true paradigm change in the sense of Kuhn.[44]

A second historical lacuna is created by the authors' insistence on the 'objective' character of their evaluations: 'objectively' Copernicus (Einstein, the phenomenologists, etc.) progressed, hence it was 'rational' to follow their lead. (I must again repeat that the standards of the methodology of research programmes are not strong enough to permit us to make such a judgement of rationality; let us forget this drawback now that we are dealing with another difficulty.) But we have seen that 'correct' actions ('correct' in the sense of the methodology of research programmes) are only rarely carried out for the 'correct' reasons. It is therefore quite possible that the Copernicans, wanting to do one thing, ended up doing another. (Newton wanted to achieve certainty, what he did achieve was unceasing progress.) In this case they acted rationally (in the sense of the methodology of research programmes) for irrational (again in the sense of the methodology of research programmes) reasons and their rationality was a matter of luck, or accident, or the propitious collaboration of irrational causes (Hegel

[42] See *De Revolutionibus*, I, 5–8, with Birkenmaijer's comments in n.82ff of G. Klaus (ed.), *Copernicus ueber Kreisbewegung* (Berlin, 1959). Cf. also Zahar's essay (254).

[43] 'Copernicus was not aware of his own riches' writes Kepler in the *Mysterium Cosmographicum*, ch. 1, n.4, by which he means that Copernicus was not aware (*a*) that he was dealing with a research programme rather than with a single theory and (*b*) that this research programme was capable of producing novel predictions. One novel prediction, mentioned by Kepler, is that the synodic anomaly of the planets depends on the *true* motion of the sun not on its mean motion as had been assumed by Ptolemy.

[44] The Aristotelian methodology of finding a point of view that could accommodate facts by degenerating absorption was not the only alternative in existence. Neoplatonists and reformers of magic such as Agrippa von Nettesheim emphasized the *hidden virtues* of objects that were not accessible to normal observation and had to be brought forth by special methods. In our period they may well have been the only thinkers to come close to the methodology of scientific research programmes. Cf. P. H. Kocher, *Science and Religion in Elizabethan England* (New York, 1953).

called this the *List der Vernunft*). The authors, concentrating on the internal features of a perhaps quite fortuitous result, leave such possibilities unexamined and thereby add to the false impression of the overwhelming 'rationality' of modern science.

6. *Judgements of progress and of degeneration are often arbitrary, for they depend on arbitrary selection of the research programmes to be compared. Altogether the appraisals made by the authors are arbitrary in at least five different ways. Once more it becomes clear that science needs and uses a plurality of standards and that scientists work best without any authority, the authority of 'reason' included*

In section 3 I argued that the standards of the methodology of research programmes are not strong enough to recommend particular actions or to characterize such actions as being rational, or irrational. We may describe the heuristic of a research programme, we may say that the programme progresses, or degenerates; we cannot infer that it may be retained, or that it must be abandoned, or that continued work on it is rational, or irrational. It was then assumed that at least the judgements of progress and degeneration themselves are made in an unambiguous and objective manner. This is not always the case.

To see this, consider two rival research programmes, *R*, and *R'*. According to the methodology of research programmes both *R* and *R'* are plagued by anomalies. Let *r* and *r'* be corresponding subprogrammes of *R* and *R'* and let the anomalies of *R* and *R'* be so distributed that *r* progresses while *r'* degenerates. A follower of the methodology of research programmes will then support *r* over *r'*. It seems that the authors of some of the papers are in this situation prepared to support *R* over *R'* as well. But in doing so they may go against their own principles for it is quite possible that *R'* progresses while *r'* degenerates and *r* progresses.

As an example, let *R'* be the Aristotelian cosmology, *r'* Ptolemaic astronomy, *r* the astronomy of Copernicus and *R* a dynamical research programme consisting of *r* and suitable dynamical principles. Lakatos and Zahar have shown that *r* progresses while *r'* degenerates. Now they are either instrumentalists or realists. In the first case they will remain content with what they have shown. In the second case they will regard the motions of *r* as real motions and infer that *R'* must be abandoned. But *R'* continued to progress long after Copernicus, as can be seen from the work of Harvey.[45]

[45] See W. Pagel, *William Harvey's Biological Ideas* (Basle, 1967) and 'William Harvey Revisited', *Hist. Sci.* 8 (1969), 1–31. Cf. also C. B. Schmitt, 'Towards a Reassessment of Renaissance Aristotelianism', *Hist. Sci.* 11 (1973), 159–93. Schmitt's problem is: what is the reason for the 'dogged persistence of the Aristotelian tradition in the sixteenth and seventeenth centuries?' (171). His answer: the ability of the system 'to adapt itself and to absorb within itself many novel elements' (178). The success of such absorption in the eyes of the contemporaries indicates that the methodology of research programmes was not universally taken as a basis of evaluation. In addition there were progressive developments.

Similar remarks apply to Zahar's comparison of the research programmes of Lorentz and Einstein. We have to distinguish L and E, the research programmes of Lorentz and Einstein, L' and E', the programmes that are compared by Zahar, L^t and E^t, the *theories* that are usually regarded as the decisive rivals in 1905, and L'' and E'', those parts of the programmes that deal with inertial spacetime only. L^t and E^t are often asserted to be equivalent, and Zahar repeats that judgement (250). He is mistaken,[46] but the mistake does not matter. Not theories, but research programmes are decisive. Lorentz's programme is said to consist of Maxwell's equations, Newton's laws of motion (with Galilean transformations) and the Lorentz force (215). Zahar never mentions the hard core of Einstein's programme though he mentions a programme, E'', that contains the relativity principle together with the principle of the constancy of c (245). In 1905 this programme started degenerating while Lorentz's programme L' was advancing,[47] and had been advancing for quite some time. Now Zahar wants to explain, in a rational way, 'why . . . brilliant mathematicians and physicists like Minkowski and Planck abandon[ed] the classical programme in order to work on Special Relativity' (251) and he also wants to show that the success of general relativity in the explanation of the perihelion of Mercury was 'a success for the *whole* relativistic programme' (262). To achieve both aims he replaces E'' by E' in the following manner (248 n.94): c is used not for 'empirical' reasons (it is not used in a degenerating way), but because of Einstein's belief in (*a*) the basic nature of Maxwell's equations (247) and (*b*) their limited validity (248, n.94). The principle of the constancy of light velocity is all that can be salvaged of (*a*) in view of (*b*), which means that it is a fundamental principle not only because of its position in the theory, but because of the nature of things. Thus E'' no longer degenerates, but it does not seem to advance either. To explain its acceptance by Planck, Minkowski and others in an 'internal' way, i.e. without recourse to psycho-sociology, Zahar turns to heuristics. To preserve continuity with the general theory of relativity and to refute Whittacker's conjecture that the ether programme was 'developed, into the relativity programme' (252), he concentrates on this heuristic to the exclusion of any hard core (265). This brings him very close to E, the always unknown research programme underlying all of Einstein's activities and thus he seems to move in the right direction. But E is still compared with L', the *truncated and frozen* programme of Lorentz, and L, which contains

[46] The Lorentz contraction involves real forces and should therefore lead to oscillations. No oscillations are to be expected on Einstein's account. No oscillations were found in the experiment of Wood, Tomlinson & Essen, *Proc. R. Soc. Lond.* A 158 (1937), 606.

[47] The light principle 'is thrown out with no justification whatever' (245), while Lorentz 'explained Michelson's result in a non-*ad hoc* way . . . and he explained the invariance of c' (236). Also 'there was no build up of unsolved anomalies which Einstein's theory dissolved better than Lorentz's' (251).

atomism as well as the possibility of a more fundamental explanation of electromagnetic phenomena is never considered. But L can take care of all the facts which E'' obtains from the relativistic formulation of the electrodynamics of media.[48] It yields the constancy of c as a contingent fact and is in this respect closer to the general theory of relativity than E'' and E' where the constancy of c is a basic law.[49] And its heuristic are at least as powerful as the heuristics of E'', for every law produced by a research programme can, of course, be used in the heuristics of that research programme.[50] We see that the choice of research programmes and rivals is fairly arbitrary, and so are the judgements that rest on them. But these judgements are the basis for Zahar's 'objective' or 'internal' appraisal of the actions of Planck, Minkowski and others.

Such appraisals (and the corresponding appraisals in the other essays of the Howson volume) therefore turn out to be arbitrary, or subjective, or 'irrational' in at least four different ways. They are arbitrary, because they proceed from an arbitrarily chosen authority: science of 'the last two centuries' (see section 2). Science is chosen not because its excellence has been shown by argument, but because everybody is impressed by it.[51] They are arbitrary because it is not really science that decides the issue – science is much too chaotic for that – but a streamlined image of it, and there are no independent arguments for the principles of streamlining chosen. Thirdly, the appraisals are arbitrary because the standards that are obtained via steps one and two are not strong enough to support judgements of rationality, or irrationality. Any such judgement is independent of the standards, it receives no authority from the standards, there are no other arguments in its favour, and so it is arbitrary, or subjective in a very strong sense of the word. Fourthly, the appraisals are arbitrary because they rest on an arbitrary choice of rival research programmes. In the case of Zahar we find yet another source of arbitrariness, and it is rather amusing: Planck and Minkowski start working on relativity, Planck and Minkowski are great scientists, hence their actions must be explained in an 'objective' manner. But there were many great scientists who either rejected the theory, or did

[48] $E = mc^2$ as well as the one sided nature of electromagnetic emission was obtained by Poinçaré in 1900 without invoking the relativistic point of view. See *Archs Néerland*, 5 (1900), 252. Hasenöhrl arrived at a more restricted result four years later. It is quite true that Lorentz himself gives no indication 'that the rest mass is a variable quantity' (262), but we are not talking about Lorentz, we are talking about his research programme.

[49] Cf. Einstein's comparison between 'constructive theories' such as the theory of Lorentz and 'theories of principle' such as the special theory of relativity in his autobiographical notes in P. A. Schilpp (ed.), *Albert Einstein: Philosopher-Scientist* (Evanston, 1951), 53.

[50] Thus the derivation on pp. 259ff of Elie Zahar's essay might have been carried out by a Lorentzian, though he would have given a very different interpretation to its result. Planck himself, who carried out the derivation, always spoke of the 'Lorentz–Einstein theory'.

[51] 'Is it not *hubris* to try to impose some *a priori* philosophy of science on the most advanced sciences? . . . I think it is.' See above, p. 210.

not pay any attention to it. As a matter of fact, 'It was only in Germany that the theory was elaborated upon.'[52] How are the actions of these dissenters to be explained? Are they to be explained in the same way as the acceptance of the theory by Planck and Minkowski? Not likely. Are they to be explained 'internally'? This would mean that a theory has advantages as well as disadvantages, that different people look at it in different ways and come to different results though they use the same set of standards. It does not seem that Zahar would accept such an explanation. But then the only way out is to admit that the dissenters acted 'irrationally', and for external reasons. But if *they* can act irrationally, then why not Planck and Minkowski? This is the fifth type of arbitrariness found in the case studies. It is surprising to see that a philosophy that makes such a fuss about 'rationality' and 'objectivity' should possess so abjectly little of either.

Zahar tries to show that Planck and Minkowski acted rationally when deciding to work on Einstein's programme and is thus faced with the problem of the rationality of those who stayed either with Lorentz, or with ether models. He neither states the problem, nor does he indicate how he would solve it. Peter Clark, in his essay on the kinetic theory, perceives an analogous problem. Towards the turn of the century the kinetic theory 'was subject to severe attacks from some of the leading scientists of the day' (42). This is not supposed to be due to purely philosophical preferences, such as a preference for positivism. All these are 'external explanations' (43).[53] The correct explanation, according to Clark, is that 'in the last decades of the nineteenth century [the kinetic research programme] was *degenerating* [while] from the two laws of thermodynamics a number of startling novel facts were deduced' (91). Was it therefore irrational to work on the kinetic theory? Not at all! The kinetic theory had a heuristic, there are means 'of systematically *improving* the theory' (75) while the phenomenological theory 'lacks a heuristic'. It is therefore also 'rational' (90) to try to make the latter supersede the former.

Now it would seem that such an attempt is rational only if a heuristic is *needed*. The kinetic theory which ran into one difficulty after another definitely needed a heuristic. The phenomenological theory which was applied to an ever-increasing domain of problems without ever failing to live up to its promise (this is Clark's description of the situation, not mine!) does not. It is too good to be in need of becoming a research programme. Besides, it was this very universality of the theory, this absence of models that made it so attractive to Einstein. Einstein[54] distinguished between 'constructive theories' which move through various stages and gradually conquer one

[52] S. Goldberg, 'In Defense of Ether' in *Historical Studies in the Physical Sciences*, ed. R. McCormmach (Philadelphia, 1970) II, 97.

[53] This definition of 'external' makes, of course, many of Einstein's reasons external.

[54] Cf. n.49.

problem after another and 'theories of principle' which remain valid no matter how far they are extended, and he preferred the latter, mentioning thermodynamics as an outstanding example. He viewed his own theory of relativity as a theory of principle, not as a constructive theory. We see that Clark does not succeed in explaining the rationality *both* of those who objected to the kinetic theory and of those who continued working on it. *At least one party must be criticized as being irrational.* Given the methodology of research programmes it is arbitrary which party we choose.[55]

This result can be generalized. Research programmes are superseded by other research programmes only because people are not invariably impressed by success and progressiveness. They are not impressed by success either because they have standards different from those accepted by the proponents of the 'successful' programme, or because they are moved by external considerations. In the first case we have a plurality of *standards*, in the second case a plurality of *motives*. Science as described in the essays reviewed here cannot exist without such a plurality and where it seems to succeed it does so only because certain problems have been overlooked (the problem of the opponents of relativity in the case of Zahar) or because of the biased terminology (talk of 'rationality' or 'objectivity' when no rational grounds for such talk have been given). The methodology of research programmes most certainly has led to some interesting historical discoveries. This is not surprising. Any hypothesis, however implausible, can widen our horizon. It has not led to a better understanding of science and it is even a hindrance to such an understanding because of its habit of beclouding facts with sermons and moralizing phrases.

[55] Peter Clark no longer believes the phenomenological theory to be without any heuristic. But, he says 'that heuristic was a weak one in a very specific sense, namely that it was *fact dependent* in much the same way that the Ptolemaic heuristic was' (letter of 26 March 1975).

11

More clothes from the emperor's bargain basement

A review of Laudan's *Progress and its Problems*

1. LAUDAN'S MODEL

In his book *Progress and its Problems*[1] Laudan presents a model of rationality that is wide enough to cover '*all* intellectual disciplines' (13), but he explains it via a discussion of its 'most successful instance', science (13). The model is simple and apparently quite powerful. It is one of those ideas which at first sight seem hardly worth a glance (11) but which reveal their fertility when developed in detail.

According to Laudan (scientific) knowledge arises from the attempt to *solve problems* (13): 'science is essentially a problem solving activity' (11; cf. 66). This is the basic postulate. It is explained by showing which entities are involved in problem solving and how the solutions are evaluated.

Problems are solved with the help of theories and *research traditions* which are sets of 'general assumptions about the entities and processes in a domain of study and about the appropriate methods to be used for investigating the problems and constructing the theories in that domain' (81, original italics omitted).

Theories and research traditions are *evaluated* by their problem-solving propensities (14). The evaluation is *comparative* (71) 'what matters is not, in some absolute sense, how effective . . . a tradition or theory is but, rather, how its effectiveness . . . compares with its competitors' (120): one chooses 'the theory (or research tradition) with the highest problem solving adequacy' (109).

In solving problems scientists 'need not and generally do not consider matters of *truth and falsity*' (24, my italics) and wisely so, for the problem-solving model works, while truth models, partial truth models and probability models don't (127f).[2] Combining theory choice with problem-solving adequacy entails that 'rationality is parasitic upon progressiveness' (125); it is not that there are two ideas, reason and progress, and we need to show how they are connected; there is just one idea of rationality where being rational already means making choices that are progressive (125).

[1] Berkeley, 1977. Numbers in parentheses refer to pages in Laudan's book.

[2] The fact that truth is absent from the *general* standards of rationality does not preclude its appearance among the *specific* parameters of particular research traditions, such as that of Kepler. Cf. 126.

The model splits rationality into two parts, a general framework that is said to be present in all cases of rational inquiry, and 'specific parameters' which are time and culture dependent (130). 'The model argues that there are certain very general characteristics of a theory of rationality which are *trans temporal* and *trans cultural*, which are as applicable to Presocratic thought or the development of ideas in the Middle Ages as they are to the more recent history of science. On the other hand the model also insists that what is specifically rational in the past is partly a function of time and place and context' (130f). This dual aspect of rationality combines the apirations of the philosopher who defends eternal rules of reason, and the relativism of the historian who asserts that reason depends on time and context. This gives us a *general outline* of Laudan's model or rationality theory.

The presentation of the general outline is combined with the discussion of *specific features*. Laudan distinguishes between empirical problems and conceptual problems and emphasizes the importance of the latter. *Empirical problems* become a challenge to a theory or a research tradition (they become 'anomalies') only when they have already been solved by some theory or research tradition (18; 29; cf. 21f: 'in appraising the relative merit of theories the class of unsolved problems is altogether irrelevant. What matters for the purpose of theory evaluation are only those problems which have been solved . . . by *some* known theory'). Being '*specific* parameters' (120) of rationality (see the beginning of the last paragraph) the conditions for the solution of empirical problems 'have evolved' (25) and they are occasionally quite loose when viewed from the standpoint of a logic freak (24). 'Assessing the importance of . . . anomalous problems for a theory has (therefore) to be done within the context of other competing theories in the domain' (38) and 'the importance of solving all empirical problems is not the same, some being of much greater weight than others' (40). *Conceptual problems* may be internal (consistency, ambiguity, circularity, 49) or external (boundary conditions such as the condition of circularity in ancient astronomy up to and including Copernicus, Einstein's 'reality condition', more general ideas such as the idea of causality and so on) and they involve theories, research traditions, entire world views (61) as well as norms: 'every historical epoch exhibits one or more dominant normative images of science. It would be a serious mistake to imagine, as many historians do, that these norms are just the concern of the professional philosopher or logician' (58; cf. 164ff for details). Laudan stresses that tensions between *world views*, theories, norms which are either overlooked or pushed aside as irrelevant by positivists (*a*) have influenced science and (*b*) have influenced it in a rational manner, i.e. in accordance with the problem-solving model: 'the overall problem solving effectiveness of a *theory* is determined by assessing the number and importance of the empirical problems which the theory solves and deducing therefrom the number and importance of the conceptual problems which the theory generates' (68, original italics). *Successful research traditions* lead, 'via [their] component theories to the adequate solution of an increasing range of empirical and conceptual problems' (82; cf. 108f for details and 119f for a summary).

The model is not restricted to science. Metaphysics, theology, even the 'formal' sciences contain empirical problems (189ff, esp. top of 191): 'What has stood in the way of a recognition of the cognitive parity of the sciences and the non sciences has been a simplistic identification of [scientific] rationality with experimental control and quantitative precision' (191); 'if there is any truth at all in the [positivistic] claim about the difference between the sciences and the non sciences . . . it will be found, not in the exclusive exhibition of progress by the sciences, but rather in the higher rate of progress exhibited by them' (192). 'Immature science', i.e. science that depends on theoretical considerations and world views, is science in the full sense of the word (155f).

The book also contains a highly critical analysis of the sociology of knowledge and it concludes with a brief attempt at judging science as a whole.

2. RELATION TO OTHER VIEWS

To judge Laudan's theory of rationality we have to examine its relation to other philosophies and its effectiveness. Where and in what respect has Laudan changed and perhaps even transcended current views; how and to what extent have the changes improved the situation?

Laudan gives us extremely bad guidance on the first question. He emphasizes the importance of a comparative evaluation of research traditions but when he comes to his own model a shoddy account of the alternatives seems to suffice. His arguments against them often have the following interesting pattern: a philosopher (historian) is introduced as holding a view, or making a suggestion *S*. *S* is examined, demolished and replaced by *Q* which is shown to be a natural consequence of Laudan's model. The model obviously is vastly better than its alternatives. Yet the poor philosopher never proposed *S*; he held *Q*, the very ideas that Laudan presents as his own. On such occasions – and they occur rather frequently – Laudan sounds like a thief who chides his victims for lacking the items he has just taken from them. This is a very clever ruse and one would like to congratulate Laudan on it but unfortunately he has borrowed it from Lakatos:[3] Laudan's 'eclecti(cism)' (ix) is much greater than he is willing to admit. A few examples will show what I mean.

The general framework of Laudan's philosophy, *the problem-solving model*,

[3] See my *AM*, 48, n.2, lines 10ff.

is of course well known (11). Anyone who has spent even a few days with Popperians and has tried to explain his ideas to them no doubt remembers the frustration caused by interruptions such as: what is your problem? You don't seem to have a problem, so what are you talking about? I don't understand your problem, so there is no use going on with your story; and so on and so forth. A lover of Platonic imagery might describe the philosophy department at the London School of Economics as a place where even the dogs no longer merely observe and react to the products of their fellow dogs but want to know the *problems* that made them produce such terrific solutions. Laudan's 'to write about the history of conceptual systems without ceaselessly identifying the problems which motivated those systems is drastically to misconstrue the nature of cognitive activity' (175) is a triviality for a Popperian as can be seen from the historical work that has emerged from that school.[4]

Problem solving can be combined with a variety of ideas about possible solutions and their evaluation. Within the Popperian circle we have the idea that solutions are proposed in the form of conjectures and are then criticized in accordance with standards which are themselves (temporary) results of a critical discussion. A theory or a research programme is judged by its problem-solving capacity, i.e. on the basis of questions such as 'Does it solve the problem? Does it solve it better than other theories? Has it perhaps merely shifted the problem? Is the solution simple? Is it fruitful? Does it perhaps contradict . . . philosophical theories needed to solve other problems?'[5] and so on. Note that these are precisely the questions raised by what Laudan calls his 'own' model (109), that they invite us to compare theories instead of trying to evaluate them absolutely and that they consider conceptual problems which, according to Laudan, Popper (and Lakatos, and I) 'simply fail to come to terms with' (47). 'What we call "science"' writes Popper[6] 'is differentiated from the older myths not by being something different from a myth' – it does not cease to contain the 'conceptual' and 'world view' (61) assumptions characteristic of mythical thinking – 'but by being accompanied by a second order tradition – that of critically discussing the myth': science is world views, etc. *plus* the problem-solving model. Popper accordingly deals with conceptual problems of early science,[7] he shows how the problem-solving model treats non-empirical questions (metaphysical theories, though irrefutable, can be evaluated by comparing their problem-solving capacities),[8] he criticizes modern physicists for failing to take conceptual problems seriously[9] and himself proposes

[4] See chs. 2 and 5 of *Conjectures and Refutations* (New York, 1962), Sabra's magnificent work on the history of optics, Lakatos' *Proofs and Refutations* (Cambridge, 1978) as well as some of the case studies in *Method and Appraisal in the Physical Sciences*, ed. C. Howson (Cambridge, 1976) and *Method and Appraisal in Economics*, ed. Latsis (Cambridge, 1976).

[5] *Conjectures and Refutations*, 199. [6] *Ibid.*, 127. [7] *Ibid.*, ch. 5.

[8] *Ibid.*, 199. [9] *Ibid.*, ch. 3 and various essays on the quantum theory.

solutions for them.[10] It is true that Popperians have combined the problem-solving model with truth, verisimilitude and corroboration and have more recently almost buried it under these ideas, but this does not impair the usefulness of the model itself for it can be discussed and developed 'without ever speaking about the truth of its theories'.[11] Laudan's '*no* major contemporary philosophy of science allows . . . for conceptual problems' (66, original italics) is therefore somewhat inaccurate, to put it mildly.[12]

[10] Example: the propensity theory of probability.

[11] *Conjectures and Refutations*, 223. Laudan even repeats details of Popper's view. Popper: 'the rational . . . character of science would vanish if it ceased to progress' *Conjectures and Refutations*, 240; Laudan: 'rationality is parasitic upon progressiveness' (125). Popper: 'the growth of scientific knowledge may be said to be the growth of ordinary human knowledge *writ large*' (*ibid.*, 216 with reference to the preface of the *Logik der Forschung* (Vienna, 1936)); Laudan: 'if there is any truth at all in the (positivistic) claim about the difference between the sciences and the non sciences . . . it will be found, not in the exclusive exhibition of progress by the sciences, but rather in the higher rate of progress exhibited by them' (192). And so on.

[12] Laudan continues: 'even those philosophers who claim to take the actual evolution of science seriously (e.g. Lakatos, Kuhn, Feyerabend and Hanson) have made no serious concessions to the non empirical dimensions of scientific debate'. But *Hanson*, like the good Wittgensteinian he was, showed the strong influence of concepts on observation and experimental matters thus turning empirical problems into conceptual problems. Summing up a series of most interesting conceptual investigations he writes: '. . . we have tried to explore the geography of some dimly lit passages along which physicists have moved from surprising, anomalous data to a theory which might explain those data. We have discussed obstacles which litter these passages. They are rarely of a direct observational or experimental variety, but always reveal conceptual elements . . .' *Patterns of Discovery* (Cambridge, 1958), 157. *The Concept of the Positron* (Cambridge, 1963) contains the following assertions: (1) 'The discovery of the positive electron was a discovery of three different particles' (135); (2) there existed a 'profound resistance' against accepting a positively charged electron (159); (3) this resistance was due to the *conceptual* structure of 'electrodynamics and elementary particle theory' (159). Can there be a clearer refutation of Laudan's complaint? Kuhn discusses a great variety of conceptual problems, both in *The Copernican Revolution* (New York, 1957), esp. ch. 4, and in his *Structure of Scientific Revolutions* (Princeton, 1962), 67, 73ff and passim. True, he makes special assumptions about the way in which conceptual problems are solved – they are developed until they generate empirical problems and then contribute to the anomalies of the underlying paradigm – but this does not mean that he disregards them or does not take them seriously (babies are taken seriously even by those who assert that they will eventually grow up). Lakatos has made us aware of long stretches of scientific development that are entirely conceptual and disregard empirical results (*The Methodology of Scientific Research Programmes*: *Philosophical Papers* I, ed. J. Worrall & G. Currie (Cambridge, 1978), 50) thus establishing the '*relative autonomy of theoretical science*' *ibid.*, 52, original italics) while his truly miraculous studies in the history of mathematics contain the best and most detailed presentation and analysis of conceptual problems in the entire history of ideas; there is nobody who has outdone him in this respect. (The objection that mathematics is not an empirical science and that Laudan's criticism applies to Lakatos' account of the empirical sciences only is removed by Laudan himself who praises Lakatos for having shown that 'even . . . the formal sciences' are full of empirical problems and therefore not essentially different from the empirical sciences, 191.) *I myself* have frequently been criticized for turning empirical problems into conceptual problems and thus robbing science of its empirical content and, indeed, most of my studies of the quantum theory, of classical mechanics (e.g. Brownian motion), of the Copernican revolution dealt with conceptual problems, problems of changing methods included. Finally, even a child is by now familiar with the way in which *logical empiricists* dwell on consistency, ambiguity, circularity, adhocness all of which are conceptual problems, according to Laudan's own classification. Result:

The same is true of Laudan's account of *research traditions*. He criticizes Kuhn and Lakatos but what he finally comes up with is hardly distinguishable from their ideas. His criticism also shows an amazing inability to understand relatively simple historical arguments. Repeating familiar complaints he calls paradigms 'obscure and opaque' (74), 'difficult to characterise' (73), 'always implicit, never fully articulated' (75), he points out that 'Kuhn never really resolves the crucial question of the relationship between a paradigm and its constituent theories' (74) and that he does not indicate at what point anomalies are supposed to precipitate a crisis (74).

Now, first of all these complaints are *not correct*. The 'difficulties of characterization' have been overcome[13] and the question of crisis is answered by Kuhn himself who points out, in perfect anticipation of what Laudan has to say on the matter (18, 21, 29) that 'every problem that normal science sees as a puzzle can be seen, from another viewpoint, as a counterinstance and thus as a source of crisis'.[14]

But the complaints are *not reasonable* either. Rationalists assume a close correspondence between science and certain basic laws of abstract thought. Under this assumption they feel justified to demand an account of science that agrees with the laws and so they ask for clear definitions, full descriptions, unambiguous rules of procedure. Obscurity and opaqueness, indecision concerning the relation between basic entities (e.g. theories and paradigms), lack of advice concerning the transition from anomaly to crisis are serious objections. They show that the analysis of science has stopped prematurely. Laudan wants a philosophy of science that is closer to the 'actual past of science' (158) and he also wants to separate eternal and specific parameters (130). This means that historical research and not rationalist declarations must now determine the nature of the entities used, their properties, their relations and their employment in the face of problems and that a *general* theory of science *must make room* for these specific parameters. It must leave specific questions unanswered and it must refrain from premature and research-independent attempts to make concepts 'precise'. Kuhn's account perfectly agrees with these desiderata. His paradigms are 'obscure and opaque' not because he has failed in his analysis but because the articulation changes from case to case. The relation between theories and paradigms remains unresolved because each research tra-

Laudan's criticism at the beginning of this footnote shows that he fails even at a very simple task, viz. the correct presentation of the views of those he had the 'good fortune' to meet as a 'student or colleague' (ix). To what extent shall we be able to trust him, when he guides us through the murkier regions of history?

[13] Cf. the work of Sneed and Stegmueller as reported in section 4 of my review 'Changing Patterns of Reconstruction' *Brit. J. Phil. Sci.*, 28 (1977). In sections 4 and 5 I also explain how the relatively stable parts of paradigms and research programmes can change and thereby refute Laudan's charge of 'rigidity' (75, against Kuhn; 78, against Lakatos). The charge is absurd in any case as the historical work of the Lakatos school shows.

[14] *Structure of Scientific Revolutions*, 79.

dition resolves it in its own way, in accordance with the cosmological, normative, empirical elements it contains. There is little specific advice concerning the treatment of anomalies because each paradigm deals with these matters in its own way. Laudan's accusation of incompleteness (which he takes over from a host of bewildered philosophers of science who have read a few logic books but have never seen science from nearby) shows that despite his severely historical posture he still shares the rationalists' dream for clear, well-defined and history-independent conceptual schemes.[15]

Laudan's accusation of implicitness, however, shows that he seems to be unaware of some very old debates concerning the difference between history and the physical sciences. Historians (and more recently Wittgenstein) have pointed out that there are practices which proceed in a strict and regular manner but with only minimal explicit knowledge of the rules, laws and standards involved. We learn a language, including the many idiosyncrasies it permits, we learn the ability to add to these idiosyncrasies in the manner of poets (or thinkers), but most of the rules that guide us are 'implicit and never fully articulated'. Learning a language or studying the regularities of a historical period does not mean studying rules in a rational manner, it means *immersing oneself* in a practice and being guided by an intuitive ability to imitate and improvise. Some older methodologists expressed this feature by saying that a historian studies a distant culture by trying to 'understand' it while a physicist who deals with explicit abstract notions explains. Kuhn makes the highly interesting and revolutionary suggestion that *physics is a historical tradition and therefore as much in need of Verstehen as history proper*. Laudan does not notice this feature of Kuhn's theory.[16]

[15] Laudan writes 'Unless we can articulate workable criteria for choice between the larger units I call research traditions then we have neither a theory of scientific rationality nor a theory of progressive cognitive growth' (106). Precisely! And unless we can articulate workable proofs for the existence of God then we don't have a good theology either. But the question is whether there are such things as 'scientific rationality', 'progressive cognitive growth' and gods. And to answer *this* question we must do some research *using concepts that are not already adapted to the rationalists' dream* and are therefore 'obscure and opaque' and 'difficult to characterize'.

[16] Summing up his account of research traditions Laudan enumerates a series of historical events and developments which according to him can be 'rationally justified' by the theory of research traditions but not by 'any other extant model of scientific growth and progress' (122f) and he concludes that 'the theory of research traditions . . . constitutes a significant improvement on the theories of rationality now in common parlance among philosophers'. But the events can be accounted for quite easily by Polanyi, Kuhn and by myself (in *A M*). They can also be explained in Popper's two-traditional model (just try it, Larry, it is not at all difficult).

Incidentally, it should be pointed out that Laudan's distinction between 'trans temporal' etc. and 'specific' parameters (130f) is old hat for Popper, Lakatos and even Kuhn. All these authors distinguish between paradigm-dependent standards and trans paradigmatic standards (developments, in the case of Kuhn). They evaluate (describe) historical episodes by

Next comes Laudan's discussion of *incommensurability*. As he tells it 'Kuhn, Hanson and Feyerabend began to despair about the possibility of any objective yardstick for comparing different theories and suggested that theories were incommensurable and thus not open to objective comparison' (143). This suggests that we wanted to compare theories, were misguided by some feature of science into believing that a comparative evaluation was impossible and joylessly published this disagreeable consequence. A look at our work reveals an entirely different story. What we 'discovered' and tried to show was that scientific discourse *which contains detailed and highly sophisticated discussions concerning the comparative advantages of paradigms* obeys laws and standards that have only little to do with the naive models which philosophers of science have designed for that purpose. *There is* comparison, even 'objective'[17] comparison, but it is a much more complex and delicate procedure than is assumed by rationalists. Thus in my first paper on the matter[18] I claim that 'a formal account of reduction and explanation is impossible for general theories' but show how predictions can still be used for comparing theories:[19] what fails is not the process of theory comparison; it is a rather simpleminded theory of explanation. According to Kuhn 'to say that resistance [to paradigm change] is inevitable and legitimate, that paradigm change cannot be justified by proof, is not to say that no arguments are relevant or that scientists cannot change their minds.'[20] 'Prob-

asking both to what extent they agree with the standards of the time and whether they are 'rational' (conform to the general pattern of development, in the case of Kuhn). With Popper the duality is part of his two-tradition model: every idea is subjected both to the standards of mythmaking (which change from time to time and place to place) as well as to the standards of critical discussion. Lakatos provides rich inventories of heuristic rules and standards including the 'normative images' that Laudan is so concerned about (58, 164) side by side with his general criterion of progressiveness. Laudan's remark that Popper and Lakatos 'insist that we should evaluate historical episodes using *our* standards and simply ignoring the appraisals made by the relevant scientists about the rationality of what they were doing' (129) is just another instance of the gulf between his account and the actual views of the people he criticizes.

[17] None of the writers who defend 'objective' standards has explained what the word means. Laudan uses the word to *criticize* but again without explaining what lack of objectivity amounts to and why it should be feared. Popperians occasionally connect objectivity with truth (in Tarski's sense) and call comparisons 'objective' only if they are based on a comparison of truth content. Incommensurability rules out such a comparison. For a Popperian the remaining standards (*and there are many standards left*) are 'subjective' which is the reason why I call them 'subjective' in my criticism of Popperians in ch. 8, section 9, 14. Laudan takes the passage as indicating that *I myself* hold them to be 'subjective' (letter of 17 June 1976) and he assumes that I apply incommensurability to *all* means of comparison, not only to means that depend on content. But already the next few lines of 'Consolations' tell a very different story.

[18] 'Explanation, Reduction and Empiricism' in vol. 1, ch. 4.8.

[19] Vol. 1, ch. 4, p. 93f. In 'An Attempt at a Realistic Interpretation of Experience' in vol. 1, ch. 2.6 (two years before I saw the manuscript of Kuhn's book and four years before the book appeared), I 'consider' the 'objection' that basing interpretations on theoretical terms 'makes nonsense of crucial experiments' and show how we can still use them. I criticize *philosophical interpretations* of crucial experiments, I do *not* criticize the *practice*.

[20] *Structure of Scientific Revolutions*, 151.

ably the single most prevalent claim advanced by the proponents of a new paradigm is *that they can solve the problems* that have led the old one to a crisis.'[21] Compare this with Laudan's '. . . an approximate determination of the effectiveness of a research tradition can be made *within* the research tradition itself . . . we simply ask whether a research tradition has solved the problems which it set itself' (145f) read in conjunction with the assertion that paradigms may have 'joint problems which can be formulated so as to presuppose nothing which is syntactically dependent upon the specific traditions being compared' (144) and one sees that except for Laudan's longwindedness there is not the slightest difference between Laudan and Kuhn. But Laudan presents his *repetition* of Kuhn as a suggestion designed to remedy a *flaw* in Kuhn's account, which is precisely the pattern I have described at the beginning of the present section.[22] There is absolutely nothing Laudan can tell us about theory comparison and theory evaluation.

This brings me to the last item on my list. Laudan not only criticizes philosophers and historians for having neglected important features of science, he also takes them to task for their 'flagrant disregard for the actual past of science' (158) which, according to him, 'is deeply grounded in their convictions about the aims of a philosophically based history of science' (168). And he criticizes especially Lakatos for 'consciously and deliberately falsifying the historical record' (170). Now while I don't know how 'conscious' Laudan himself was when telling his fairytales about Popper, Hanson, Kuhn, Lakatos and the humble author of the present review and how 'deeply grounded' his fabrications are in his wish to appear original, I am certainly amazed at the difference between these fabrications and the 'historical record'. The accusation just quoted is another instance of this pattern. Take Lakatos; he writes[23] 'In writing a historical case study one should, I think, adopt the following procedure: (1) one gives a rational reconstruction; (2) one tries to compare this rational reconstruction with actual history and to criticize both one's rational reconstruction for lack of historicity and the actual history for lack of rationality. Thus any historical study must be preceded by a heuristic study . . .'. He illustrates the principle partly with sketches whose 'caricature'–character he explicitly emphasizes,[24] partly with detailed studies such as his incomparable *Proofs and Refutations* where the reconstruction is presented in the form of a debate

[21] 152, my italics.

[22] Thus Laudan (145f), partly using Kordig, proudly presents procedures that survive incommensurability, and implies that none of us ever thought of such a clever escape. But 'Consolations' discusses *exactly the same* procedures, and in greater detail than Laudan while Hanson, in his magnificent analysis of the correspondence principle, showed long ago how incommensurable theories can be compared and so made an important contribution to our understanding of research amidst changing ontologies (see *Patterns of Discovery*, 148f).

[23] *Philosophical Papers*, I, 52, in italics in the original.

[24] *Ibid.*, 55, n.3.

while 'the real history . . . chime(s) in the footnotes, most of which are to be taken, therefore, as an organic part of the story'.[25] Where is the 'conscious and deliberate falsification'? Lakatos' reconstructions are *blueprints* which he presents *in addition to* the buildings whose structure they are supposed to determine. Nobody would call a blueprint a 'falsification' because the builders chose to disregard it.[26] At any rate, the procedure is very different from Laudan's who introduces his account without qualifications, as if it were already the Real Thing.[27]

Laudan's description of the present situation in the philosophy of science is an extreme example of a widespread phenomenon: every profession has a body of beliefs which are hardly ever examined, are out of touch with reality and yet play an important role in arguments and the associated propaganda. Examples are the assumption of the empirical nature of the Copernican revolution, the assumption that Newton derived the law of gravitation from facts, that Boltzmann was an old fashioned realist fighting valiantly against positivist deadbeats such as Mach and Ostwald, that Marxists live off *ad hoc* hypotheses, that Einstein took falsifications seriously, that astrology has no connection with reality, that every illness proceeds from a localizable material process. Laudan's book, despites its belligerently historical stance, reveals some fairytales *in statu nascendi* and its reception shows how quickly and readily philosophic folklore accepts them.[28] I have tried to restore – not too successfully, I am sure – at least part of the real story. What remains? *Popper's original problem-solving model* freed from the cumbersome logical machinery which Popper himself and some of his more distant pupils have superimposed on it,[29] supplemented with a pinch of

[25] *Proofs and Refutations* (Cambridge, 1978), 5.

[26] This also answers McMullin's criticism mentioned by Laudan (168, n.17).

[27] To show my shortcomings Laudan refers to papers by McEvoy and Machamer (168, n.17). But McEvoy cannot be taken seriously (cf. my *SFS*, 160 and n.17) and Machamer's history, though more bulky, is hardly better than Laudan's. Besides, I have replied to him and refuted his criticism point for point (see the reprint in *AM*, 112f). Laudan, wisely though somewhat disingeniously, neglects to mention this reply even though it was published in his own journal.

[28] 'Ringing in the New' is the title of two reviews, including Burian's review of Laudan's book: *Isis*, 69 (1978), 602.

[29] It is interesting to see that fundamental discoveries which show the limitations of simple-minded modes of thought are as a rule succeeded by the belligerent reaffirmation of these modes. Popper criticized the formalist character of the Vienna-Circle philosophy but he soon introduced technicalities of his own (corroboration; verisimilitude) which for his less gifted successors have become the Alpha and Omega of rationalism. Today the problem is no longer the advancement of science but the preservation of a school philosophy. Ordinary language philosophers once laughed at the childish pretensions of formal logic only to fall for them in the end. Imre Lakatos gave splendid examples of the looseness of proof patterns in informal mathematical logic only to prefer a more rigorous logic towards the end of his life (cf. the editorial comments on p. 138, n.4* and p. 146, n.2* of *Proofs and Refutations*). Even the sceptics did not escape this pattern: the informality of Pyrrho was soon followed by the learned discourses of Carneades and completely killed in the presentation of Sextus Empiricus.

Kuhn and seasoned with generous helpings from the work of Laudan's other victims. Let us now see how this model fares when compared with science, commonsense and itself!

3. ADEQUACY OF THE MODEL

Laudan makes a point of asserting that his model, while vastly more liberal than its competitors, is not without limits. 'To suggest that "anything goes", that any combination of beliefs would emerge as rational and progressive on this model is profoundly to misunderstand the high standards of rational behaviour which it requires' (128). Moreover, 'for scientists in any culture to espouse a research tradition or a theory which is less adequate than other ones available *within* that culture is to behave irrationally' (130; adequacy is defined as problem-saving capacity, 109, 124). It is also irrational to *argue* against theories or research traditions on the basis of non-progressive world views (132). Finally, Laudan would presumably regard it as irrational to *pursue* inadequate research traditions of a (comparatively) low or even a negative ratio of progress (cf. 111). His model does indeed impose limits. But the question is: are the limits important, are they realistic, can they be upheld within the model? It seems that the answer to all three questions is no.

To start with, let us point out that a model may have content in the sense that it forbids actions and calls them irrational but may still be practically vacuous in the sense that the forbidden actions are of no interest to anyone. Lakatos, for example, does not permit people to *call* degenerating research programmes progressive – *this is the only 'rational' weapon he produces against 'chaos'* – but who is going to start a war over such a trivial matter?[30] Laudan's objection to 'espousing' inadequate theories, though less obviously trivial, has similar cash value: inadequate theories may be 'pursued', they may not be 'accepted' (180ff). What does it mean to 'accept' a theory? 'to treat it as if it were true' (108, original italics omitted). Yet according to Laudan the relation of theories to truth plays no role in science. So it cannot be used to separate acceptance from pursuit. What

[30] For the background of this criticism see ch. 10 (esp. end of n.25), which contains my comments on the essays in Howson, *Method and Appraisal in the Physical Sciencies*. A. Musgrave in *Progress and Rationality*, ed. G. Radnitzky & G. Anderson (Dordrecht, 1979), 192 objects: 'Lakatos is no *epistemological* anarchist since he provides a whole battery of standards for judging theories and research-programmes.' Now first of all it is not Lakatos who 'provides' these standards, but scientists: none of the standards which Musgrave mentions further down on the page were invented by Lakatos. He took them from history. Secondly, scientists use these standards *opportunistically*: they sometimes follow them, but at other times disregard them. Lakatos condones this opportunism pointing out that it is backed up and held together by his general theory of rationality. But this general theory does not work; even Musgrave admits this now. What remains are many rules guided by a healthy opportunism that changes from one case to the next. In other words *anything goes*.

remains is the order *not to say* that one has 'accepted' a theory that is comparatively inadequate and yet plays a role, but to speak of 'pursuit' instead. Big deal! The appeal to commitment (a scientist who has accepted a theory 'must commit himself', 109) does not help. First, because according to Laudan commitment may be tentative to a high degree ('however tentatively', 109) and, secondly, because there is no behavioural difference between commitment and vigorous pursuit: if there are various parths open to you for reaching an aim and you are not sure which path to choose you may start walking along the first path, which is exactly what you would be doing if the path were the chosen one. But assume we admit that not all theories are treated in the same way: some medicines are fed only to rats while others are released for human consumption. Then the difficulty is that this difference does not help us with purely theoretical problems; nor is it clear that we are dealing with a difference between acceptance and pursuit rather than a difference between different forms of pursuit. Some pharmacologists may of course say 'this is it!' and stop looking for side effects, but it would be more than a little absurd to honour such an attitude by creating a special epistemological category; and if the category exists despite its absurdity then it is wise, in the interest of human welfare, to stay on the side of pursuit and to warn patients against doctors who have moved over to acceptance. I conclude that the distinction between acceptance and pursuit may characterize *special cases* but it would be vacuous, or at least unwise, to make it a basis of general rules for the evaluation of research traditions.

The conditions of pursuit, on the other hand, are much too restrictive. Laudan distinguishes between the *adequacy* of a theory – it has *solved* more problems than its competitors – and its *promise* which is the *rate of progress* in solving problems. A highly inadequate theory may be promising in this sense (112f) and so deserve to be pursued, while inadequacy and lack of promise count against it: a theory, a research tradition, or a world view *must perform well* before it can become part of research. But how can we judge its performance if we have not already made it part of research? To object to the pursuit of an idea unless there is some guarantee in terms of performance is putting the cart before the horse, for the required guarantee can be obtained only by means of the very research one wants the guarantee for.[31] And indeed we find, when looking at history, that lack of performance and inadequacy have never stopped people from pursuing views they regarded as important. Atomism, Platonism, the idea that the earth moves, the idea that the laws of nature have a history, and relational accounts of space and

[31] The fact that Lakatos and Laudan introduce rules of pursuit shows that they have not understood Popper's (or, rather, Mill's) anti-inductivism (I am now thinking of Mill's *On Liberty*): justification *comes with research*, it cannot be a *precondition* of it; nor can one expect it to turn up within a well-defined time interval, as a result of steps that are known in advance.

time were proposed or revived not because they had performed well in the past but because they were believed to possess an (as yet unrealized) *ability* to perform. The many revivals of Platonism, atomism, of magical world views, the rise of rationalism in Greece can hardly be explained in Laudanian terms, even if we consider conceptual matters only. These phenomena have much in common with revivals of *faith*, but they contributed to the advancement of science. The same faith in potential rather than in actual performance, or 'promise' in Laudan's sense, was responsible for some of the most interesting developments in the history of thought. At the beginning of Western rationalism, abstract argument was faced by almost insurmountable problems (the paradoxes of Parmenides and Zeno; difficulties in mathematics; the problem of the relation between commonsense and philosophical theory, city law and philosophical law, perception and 'reality' – all of which were noticed and discussed in the works of Plato and Aristotle). Some of the problems were 'solved' (very often in an *ad hoc* manner, by turning them into principles) and the solution created more problems. This delights critical rationalists but is a difficulty for Laudan: a research tradition which in solving problems proliferates problems has a negative rate of progress; yet one continues on the troubled path and uses it to change existing traditions which are adequate and perhaps even progressive. The Copernican revolution is another instance of the principle, denied by Laudan, that in intellectual matters it is expectation, faith, hope or simply ignorance (of problems) and not actual performance that explains pursuit. In the *Commentariolus* Copernicus criticized astronomy for its reliance on the equant. The principle of his criticism was the idea that real astronomical motions are centred circular motions with constant angular velocity and that an explanation of phenomena must consist in their reduction to such motions. In 1520/40 the principle was neither adequate nor progressive in Laudan's sense while the equant was at least adequate. Yet the problems created by the conflict between the principle and Ptolemaian astronomy were taken seriously enough by Copernicus, the Wittenberg astronomers (as reported by Westman), and Brahe to justify attempts at rebuilding astronomy: theories or world views are permitted to create problems even though they are neither progressive nor adequate. Moreover, the Copernican arrangement was itself beset with problems. The problems are disregarded by philosophers of science who restrict themselves to astronomy. But expanding the domain of discussion into physics, optics, theology as Laudan suggests, and counting successes *as well as failures* (78, item 5, against Lakatos) the rate of progress is considerably decreased. In the seventeenth century the situation became more opaque, but not better. For the new philosophies that were now introduced to accommodate the new cosmos created further problems, such as the mind–body problem, the problem of the relation between God and the world, the

Word of God and the Work of God (taken very seriously by Newton), the problem of motion (in the Aristotelian sense, including qualitative change), which are fundamental, have greater weight than technical problems, and which have resisted solution to the present day. The elimination of witchcraft theories, to mention only one side effect of the development left a great variety of psychological problems unsolved without having anything better to offer; and this situation lasted until the nineteenth century. *Nobody knows* what the *overall rate of progress* was; nobody knew it then and *nobody cared* partly because the difficulties were not noticed by those 'at the forefront of research' (ignorance), partly because they were not regarded as important, partly because special achievements in a narrow domain were regarded as sufficient reasons for carrying on at all fronts (for the period in question Lakatos' emphasis on success over failure – criticized by Laudan (78) – is on the right track) but mainly because potential was more important than promise in the sense of Laudan (on this last point Lakatos and Zahar again seem to be more clearsighted: they emphasize progressiveness but refrain from turning it into a principle of pursuit; and they make an analysis of the *heuristic promise* of a research programme an essential part of its evaluation).

I have said that world views are often used in arguments against successful traditions even although they are neither adequate nor progressive and I have mentioned nineteenth-century atomism, ancient Greek rationalism and Copernicus' principle of circular motion as examples. Actually, the situation is much more complex. Principles are never used in isolation, but in conjunction with other principles. A new idea that clashes with established results and gives rise to numerous conceptual and empirical problems often gains strength from other ideas that are equally inadequate and unpromising (in Laudan's sense) but support it and gain support from its further articulation. Everybody who has taken the trouble to study the philosophical systems of Hegel or Aristotle has no doubt felt the intellectual force emanating from the collaboration of ideas which, taken one by one, are implausible, unrealistic, in conflct with their surroundings.[32] In this way Parmenides' ideas drew strength from their internal coherence although they clashed with everything around them, progressive practices included. And this is also the way in which ancient ideas such as the atomic theory survived through the ages and finally overcame their most successful rivals.

To sum up, inadequate and unpromising research traditions, world views, and theories *are* often used as parts of research; they *must be* used in

[32] Duhem has gone far in reconciling a basic Aristotelianism with the methods of modern science (Simplicius and Aristotle himself preceded him in this respect). And Professor Kuhn once told me that seeing the internal coherence of the Aristotelian philosophy and its ability to deal with problems in its own terms was an important step towards the theory of paradigms. Cf. his *Essential Tension* (Chicago, 1977), xiff.

this way if their virtues are ever to be revealed and they *are aided* in their use by the mutual support of the ideas they contain. Faith, good sense and internal articulation collaborate in a procedure which, according to Laudan, is 'patently absurd' (132). Moreover – and with this we come to a further objection against him – such use may considerably reduce the success of the theories criticized and with it the absurdity of the procedure. As Laudan presents the matter, world views, theories, research traditions are either progressive or not progressive. Only in the first case are they permitted to judge and to criticize. How do we find that a world view is non-progressive? By showing that it has many problems but has neither solved them nor shown promise of solving them. How do the problems arise? From conflicts with research traditions, facts and so on. Such conflicts can be used against the world view, but they can also be turned against the research traditions. In the first case we have a world view with problems, in the second we have a research tradition with problems. According to Laudan the direction in which the conflict is turned depends on the past history of the world view: if many problems were solved in accordance with its principles, then the conflict can be turned against the research traditions. Otherwise the world view is disregarded. We have already seen that this is an unreasonable step: it would forever exclude new world views from entering the scene. If we want to try out such views, then we must be prepared to take the 'patently absurd' step and must turn them against successful traditions even before they have achieved their own first success. We must permit them to create problems for these traditions and challenge the defenders of the traditions to solve the problems. We have also seen that this is indeed the way in which revolutionary changes are brought about; but now we have a new problem: how do world views which have received the power of life or death over research traditions ever lose this power? Or to put it differently, how does it happen that the problems they create are taken less and less seriously and are finally not regarded as problems at all? For example, how is it that the problems that a substantial Christianity created for the mechanical world view (and which Newton took very seriously) are no longer with us? Because the mechanism that guides the exchange of world views *constitutes* what counts as a problem or as a good performance and is therefore relatively independent of performance and problems. What is this mechanism? Change of allegiance to another world view or another research tradition. *Change of allegiance, i.e. a socio-psychological process, is primary; calculation of performance, 'rationality', comes after it and depends on it.* So we are finally back at our earlier result viz. that (excepting special circumstances) *performance has no direct influence on the acceptance or rejection of views* and it cannot have such an influence, by its very nature. Reason, however, turns out to be a local agent whose application and whose limitations depend on circumstances of an entirely different kind.

There is still one element missing from my evaluation of Laudan's version of Popper's model: the model has many ways of *circumventing* rules without violating them. There arises therefore the suspicion that a determined application of all the methods it offers can overcome difficulties only by reducing content in precisely the manner in which Lakatos' model, in trying to survive attacks, is finally pushed into excluding trivialities only. The suspicion is confirmed by noticing that Laudan accommodates empirical and conceptual problems with the help of *ad hoc* hypotheses (115), or by denying that the theories that give rise to conceptual problems make substantive assertions (instrumentalism), or that their observations are genuine observations (devaluation of observations by declaring them to be illusions). Laudan's model can reduce rates of progress by denying the existence of problems already solved, and increase them by either changing from one methodology to another (permitted by Laudan (59)) or from one world view to another, or else by restricting research programmes to domains where they show success and making the success achieved there a measure of their overall progressiveness. All these procedures have occurred in the history of science, they have advanced (in our sense, or in the sense of the users) science at decisive periods and they can therefore be supported by pointing to the 'actual development of science'. But the trouble is that Laudan, in making them part of a *unified* theory of rationality with *general* rules such as those quoted at the beginning of the present section, will be inclined to use them where they do not help, or to void them of content. The trouble is that like Lakatos before him Laudan gives us a model that is either inadequate or trivial.

But rationality was not always in this predicament. When rationalism arose in Greece it created mathematics, astronomy (in our sense), the history of ideas, biology, psychology, the theory and practice of drama, theology. All these subjects were created as the result of procedures which one might call rational. But the *study* of knowledge soon became separated from the process that *created* it. While Aristotle examined (the conditions of) existing knowledge and created new knowledge in accordance with the results of his examination, his successors, among them Kant, Mill, Whewell were content to explore the structure of an already existing body of knowledge without adding to it. In the twentieth century attention was concentrated on the instruments of the exploration. Logical sophistication increased, but the ongoing process of (scientific) research became more remote than ever. Lakatos and Laudan are the late children of this development. Once, long ago, Lady Reason was a beautiful, strong, helpful though somewhat overbearing Goddess of research. By now her lovers (or, should I rather say, pimps?) have turned her into a garrulous but toothless old woman.

Sources

The publishers thank the editors and copyright holders who have given permission for these essays to be reprinted here: chapter 2, 'Classical Empiricism', from *The Methodological Heritage of Newton*, ed. R. E. Butts and John W. Davis, Blackwell, 1970, pp. 150ff; chapter 3, 'The Structure of Science', *British Journal for the Philosophy of Science*, volume 16, 1964, pp. 237ff; chapter 4, 'Two Models of Epistemic Change: Mill and Hegel' was part of 'Against Method', *Minnesota Studies in the Philosophy of Science*, volume 4, 1970, pp. 27ff; chapter 5, 'Philosophy of Science versus Scientific Practice: Observations on Mach, his Followers and Opponents', as part of 'From Incompetent Professionalism to Professionalised Incompetence – the Rise of a New Breed of Intellectuals', *Philosophy of the Social Sciences*, volume 8, 1978, pp. 37ff; chapter 6, 'Mach, Einstein and the Popperians', under the title 'Zahar on Mach, Einstein and Modern Science', *British Journal for the Philosophy of Science*, volume 31, 1980, pp. 273ff; chapter 7, 'Wittgenstein's *Philosophical Investigations*', *Philosophical Review*, volume 64, 1955, pp. 449ff; chapter 9, 'Popper's *Objective Knowledge*', *Inquiry*, volume 17, 1975, pp. 475ff; chapter 11, 'More Clothes from the Emperor's Bargain Basement: a Review of Laudan's *'Progress and its Problems'*, to appear in the *British Journal for the Philosophy of Science*.

Name index

Subject index